WOMEN IN PRINTING

Northern California, 1857–1890

MRS. ELIZA G. RICHMOND

Superintendent, then proprietor, of the
Woman's Co-operative Printing Union

~~Women in~~ Printing

NORTHERN CALIFORNIA, 1857–1890

ROGER LEVENSON

CAPRA PRESS

SANTA BARBARA

Library of Congress Cataloging-in-Publication Data

Levenson, Roger, 1914–1994
    Women in printing : Northern California, 1857–1890 : including a
roster of women and a checklist of imprints / Roger Levenson.
            p.   cm.
    Includes bibliographical references.
    ISBN 0–88496–365–9 (pa.)
    ISBN 0–88496–   (cl.)
        1. Women printers—California—San Francisco Region—History—
19th century. 2. Printing industry—California—San Francisco Region—Employees—History—19th century. I. Title.
Z243.U6C2 1994
686.2'082—dc20   94–9886
                                        CIP

CAPRA PRESS
P.O. Box 2068, Santa Barbara, CA 93120

To the memory of
EMILY PITTS STEVENS
one of the "strong-minded females;"
editor, master-printer and publisher of
the first suffrage journal in the West

TABLE OF CONTENTS

CENTRAL PACIFIC RAILROAD.

No.............   SAN FRANCISCO, ... 18 80

Received, *by the* CENTRAL PACIFIC RAILROAD COMPANY, *of*

C. JAMES KING OF WM. & CO.

(who ha agreed to the conditions and agreements herein mentioned) the following property, in apparent good order, except as noted, marked and consigned as below, which they agree to deliver with as reasonable dispatch as their general business will permit, in like good order (the dangers incident to Railroad transportation, loss or damage by fire while at depot or Stations, loss or damage of combustible articles by fire while in transit, and unavoidable accidents except), at

....Station, upon the payment of charges.

Conditions and Agreements:

Whereas, it is estimated that the cost of properly box ng and packing Sewing Machines, Baby Carriages, Iron Pumps, Light Castings, Furniture of all kinds, Earthen Ware, Drain Tiling, Flowers and Plants, Agricultural Implements, Machinery, Marble, Stoves and Stove Furniture, Wagons and Carriages, and all other freight of a fragile nature, or such as is liable to injury through not being properly boxed, is greater than a reasonable risk of injury to shipment when not so packed, it is distinctly agreed and understood, that in consideration of the receipt of such articles as offered for transportation, the Railroad Company is released from all liability for damage by breakage, chafing, or other injury resulting from insecure packing, and for all loss or damage to said articles not resulting from theft, conversion of property, or from collision of trains, or from cars leaving the track.

That Oils and all Liquids are at owner's risk of leakage; that the Carrier is not responsible for the leakage of liquors in glass, or for breakage of any article packed in boxes when said boxes show no signs of rough handling.

That the Company agrees to forward the property to the place of destination as given below, but its responsibility as a Common Carrier is to cease at the Station above named, when the property is to be delivered to connecting roads or carriers.

That the Company does not agree to carry the property by any particular train, or in time for any particular market.

That Eggs, Oysters, Poultry, Dressed Hogs, Fresh Meats, and Provisions of all kinds, Trees, Shrubbery, Fruit and all perishable property is at owner's risk of frost and decay.

That as a part of this agreement, all other carriers transporting the property herein receipted for as a part of the through line, shall be entitled to the benefit of all the exceptions and conditions above mentioned, and if a carrier by water, he is entitled to the further benefit of exception from loss or damage arising from collision, and all other dangers incident to navigation.

No. Pkgs.	DESCRIPTION OF ARTICLES.	WEIGHT, Subject to Correction.
1	Peas	
3	" Beans	
3	" Tomatoes	
1	" Corn	
8		

*When freight is destined off or beyond the line of C.P.R.R., Agents will be careful to note that the name here inserted is the station at which freight leaves the road.

Agents will be particular to number both Receipt and Shipping Order, marks and articles on which must be alike.

Received, subject to above conditions and agreements.

....Agent.

Women's Print, 521 Montg'y St.

LIST OF ILLUSTRATIONS

Fig. 2 , facing page: *An original waybill printed by women for the mighty Central Pacific Railroad whose Leland Stanford was a staunch supporter of woman suffrage.* Author's Collection.

PREFACE

My friends are too well aware that this study has been germinating over an unusually long time. There were many reasons for the delays in organizing such a complex text which offers much disparate material that exists nowhere else. The Appendixes, with a roster of women in printing in San Francisco and Northern California and the large listing of their identifiable imprints, represent a vast amount of painstaking searching but are still only beginnings. I hope others may see fit to add to them. Further, the many, many newspaper references could be very useful to anyone who wishes to undertake a badly needed, extended survey of the woman's movements in San Francisco during the period of this study. The references from newspapers are only starting points for many different topics and individuals. In my opinion, several important persons mentioned briefly in the text warrant detailed treatment.

The status of women bore heavily on difficulties in getting information. If you add the earthquakes and fires, especially 1906, almost all the meager records that might have been made for a woman no longer exist. It is ironic that the whole U.S. Census of 1890 was also lost by fire so that any further checking in that invaluable source is an impossibility.

As noted elsewhere, neither the Census data nor Directories are completely reliable sources but lacking any other basic data at all, I have been forced to rely heavily upon them for information on individuals, always double-checking when it was possible. I have read the Census sheets for San Jose, San Francisco and Oakland for 1860, 1870, and 1880 and so many Directories line-by-line that I get eye strain merely writing about it. But this was the only way to make any kind of a roster which would give even a small overview of the range of persons, places and work. There were surprises, including an 11-year-old girl (A 156) "apprenticed to a printer" in the 1870 Census, but the cumulative result is more important than a few, unusual entries. (When an individual in printing is first introduced in the main text, her/his numerical listing will be given as above and will direct the reader to detailed information in the appropriate Appendix.)

In keeping with many authors, I have made some arbitrary decisions to simplify the handling of repetitive items. The paragraph immediately above is an example. Despite my reliance generally on *The Chicago Manual of Style*, I decided it would be much easier and less fussy to omit the italics for Census and Directory. Likewise, I shifted to the simple style of the LC cataloguing rules in the imprints checklist, again with no italics but with capitalization. However, within the text proper and footnotes, I have tried to adhere to basic *Chicago* practice in other matters. The introductions to Appendixes A and B note other vagaries I have indulged for simplifying those sections of the book.

The few footnotes with newspaper references which I found in books had so many "ghosts" and non-sequiturs, I decided to stick with the now old-fashioned practice of giving page and column (e.g., p. 0:0), while indicating the volume separately (e.g., vol. 2, p. 0:0). Microfilm is another headache-inducing research tool when used too long so anything that speeds up the process seemed to me worthwhile to incorporate into my manuscript.

I am sure there will be those who find this book contains too many quotations. After reflecting on how I was going to proceed, I decided that wherever possible it was better to have the individual speak for her- or himself rather than merely summarizing. Certainly Lisle Lester's vivid prose, and some of Emily Pitts Stevens's more truculent effusions make good reading and give insights into the individuals. When a recent book about editors in the West came out, I was somewhat shocked to find Lisle Lester relegated to half a sentence. I have devoted Chapter 4 to her. Another person now is attempting a full-length study of this most interesting, colorful editor and printer. The newspaper prose directed towards women and their activities often illustrated vividly some of the attitudes women were trying to combat, so I elected not to summarize very often.

Another decision that took much thought and weighing of pros and cons was to take Chapter 2, which describes typesetting in the kind of detail I do not believe exists elsewhere, and move it from the Appendix to become a self-standing chapter. I warn readers that skipping this material will diminish understanding the text that follows considerably. Elsewhere, I go into the whole confusion of the use of the word "printer" in some detail and with good reason: even reliable authorities tend to mix their terms and thereby confuse the issue — especially about what the working women really did in San Francisco from 1857–1890 in the printing trade. The Introduction to Appendix A is required reading for an understanding of the disparate job descriptions that appear there. As noted in Chapter 9, the California Commissioner of Labor Statistics was also prone to misunderstanding what different terms specifically meant in the way of work done, thereby muddling the issue in his discussion. I think the easiest way to encompass all of this looseness in using printing terms, perhaps, is to think that, with the exception of owners or bosses, all of the women in this book did nothing but typesetting in the composing room, just as many modern women have done nothing but typing in offices. Female typesetters occasionally became forewomen but the question of what they did, other than supervise, is not clear. Possibly a small few made up printing forms, thereby becoming *compositors* by definition. *Proofreading*, along with *feeding presses* appear to be the only other work regularly performed by women.

Rather than extend this work unduly, I decided to omit a glossary of printing terms. I have italicized and defined within the text the essential terminology as it occurs and have italicized certain recurrences to emphasize the special use of the word(s). Most any dictionary can help if you need further information and, of course, specialized printers' dictionaries are the best and are frequently found in public as well as research libraries. Context will often assist understanding as well.

There has been a tendency, since the resurgence of the woman's movements, to romanticize the role of women in printing. One reason locally might be the late — and truly remarkable — Jane Grabhorn, who learned to print in the family business so that, eventually, she designed and published some very outstanding books, but not because of commercial necessity. This was fun for her, fun for those who enjoy her wit and humor to this day, and she furnishes wonderful memories to those of us who knew her. But Jane Grabhorn was a notable exception. The confusion here is the growth of so-called private presses or "fine printing," terms which encompass Jane Grabhorn's work. Most of the women I know who chose this route to make a living during the past 25 years have failed, not because of lack of talent but because of economics. Hobby, and semi-hobby printing, are to the contents of this book as mangoes are to apples. However, many women have been successful in recent years with commercial ventures because of the changing technologies which have opened new opportunities to them. Even the often-cited Emily Faithfull, who started a 19th Century printing-office in England exclusively for women workers, gave it all up after a couple of years and became a full-time publisher.

Printing is hard, grubby work, especially when you stand on your feet up to ten hours a day setting type as fast as you can under what rarely were ideal working conditions (see Chapter 9). There is nothing romantic about the women in this study who were strictly typesetters. Unfortunately, many of their battles in the workplace are still being fought.

I am obliged to so many people for assistance over the past thirty years that there is no way I can adequately list or thank all of them. The project started with the late Freda (Mrs. Lawton R.) Kennedy, Oakland, who first brought the matter of women in printing to my attention. Miss Reda Davis, Pacific Grove, then kindly shared many notes on the woman's movements — her specialty — and helped me to identify persons of importance in this book about whom I knew nothing when I started. Historian Richard H. Dillon has been a faithful donor of references since 1958. The staffs of The Huntington Library, San Marino; The Bancroft Library, University of California, Berkeley; the California Room of the California State Library, Sacramento; and Special Collections, San Francisco Public Library, have been especially helpful over a long period of time and my gratitude knows no bounds. The Heatherly Family, Oakland, direct descendants of Mrs. Lizzie G. Richmond (Chapter 7), was more than generous in sharing information and especially the rare family photographs to enhance this book, and I thank them for everything. Michael Engh, S.J., Los Angeles, has been most obliging with data concerning his great-great-grandmother, Nellie Gorman, and in furnishing photographs.

Many quotations are also due to the generosity of others, especially Dr. Robert J. Chandler, an historian at the Wells, Fargo Museum, San Francisco. He wrote his Ph.D. thesis using primarily California newspapers from 1858 to 1868. As he canvassed for his political interests, he soon found many references to women which he noted and some of which he later used in a published article about Lisle Lester. His sharing these leads led to some great additions to this study, including the physical description of Lisle Lester herself. The *Los Angeles Star* citation about the young person who became their first female typesetter, and probably the first in that area as well, is a distinct contribution.

My long-time friend, Allan R. Ottley, Sacramento, contributed innumerable citations that were most useful. One of the most recent is the best; neither of us had an inkling of what the article would contain. It is Allan Ottley we can thank for the wonderful story about Kitty Burke (A 36), San Francisco, and her typesetting feats (with drawings, too) in Chapter 2.

Thank you again, Robert J. Chandler, for your valuable assistance in editing this manuscript. Edwin H. Carpenter, Pasadena, and the late James DeT. Abajian, San Francisco, also have been helpful. Most importantly, the editorial and technical assistance of Judith A. Sutcliffe, Santa Barbara, have made this book possible. I am also indebted to Richard N. Beatty for other technical help. I am grateful to another longtime friend and colleague, Prof. Robert D. Harlan, San Francisco, for counsel and especially for his contribution of so many entries for the Checklist of Imprints from his ongoing survey of early San Francisco printing. All have most willingly shared and that has been a significant part of the pleasure of writing this book.

Roger Levenson
Greenbrae, California
7 February 1994

(*Roger Levenson died on May 29, 1994.*)

Chapter 1   Women in Printing History

Women have been involved with printing since its beginnings in the mid-15th Century. Countless anonymous women — and children — cohabited where printing was a household enterprise until the early part of the 19th Century. A person living with a printing-office underfoot was exposed daily to its diverse routines but, more importantly, had an extra set of hands that easily could be put to use when required. As a consequence, using the basic term, "printer," to describe the variety of work women performed is imprecise and gives no clue to the division of labor that always existed in the printing trade.

Both the printed record and day-to-day speech contribute to the problem. For example, "Printed by D. B. Updike, The Merrymount Press," is a widely known American imprint. However, Updike reiterated — truthfully — in both his writings and lectures, that he never "printed" anything. In fact, he was one of the greatest master-printers, "…the soul of printing; and all the workmen as members of the body governed by that soul subservient to him …" as Joseph Moxon colorfully defined the term in 1683.[1] In short, he was the boss, the person with the ultimate responsibility for the printing-office. Many women of record have been able to assume this role in their deceased spouses' businesses successfully during the years of household enterprise. That fact alone proves that these women had learned much about printing, yet the history of their *specific* experiences is frustratingly sparse. The first illustration of a woman setting type, with the *composing stick* held correctly in her left hand, is in a printer's device of Badius Ascencius and was used on the title page of a book printed in 1523.[2] However, women were setting type long before the time of Ascencius.

The earliest documented women in printing were nuns. Updike says that the Ripoli Press ". . . furnishes an early instance of the employment of women in composing-rooms [before 1480]."[3] However, nuns were atypical because their special status within the religious community contrasted sharply with that of women in the society around them. Nuns were often educated and trained for specific work so that many were already associated with bookmaking, before the advent of printing, as scribes, illuminators and rubricators. Their appearance in a printing-office early in the history of the craft may have been a concomitant of this previous connection with manuscript books. It is very likely that their special status as educated and trained women brought them into the religious-oriented Ripoli Press and similar establishments. One standard reference quotes a report that the nuns may have also done presswork. "From 1476, some eighty books were issued from this press."[4] (Signed bindings by nuns of the 15th Century are also well known.)

While data on the Ripoli nuns contain uncommon specific references to tasks performed, until the 19th Century almost all of the known women turn out to have been, in fact, master-printers — the persons who owned the businesses, or managed them for another, and who normally would have directed all activities personally. However, "Estellina Conat, the widow of a printer," states, in a Hebrew book (Mantua, before

1480), that "she and one man did the typesetting," clear testimony that she was prepared to carry on both the technical and supervising aspects of the firm.[5]

The first printing-office independently owned and operated by a woman is generally agreed to have been that of Ann Rügerin, who was "printing" in Augsberg, Germany, in 1484. A needless confusion appears when Colin Clair says that the *Sachsenspiegel* of Eike von Repgow was printed at "the first press owned and worked by a woman."[6] Press, as used here, is commonly synonymous, on the one hand, with a printing-office, and a printing-machine, on the other. The word "worked" means specifically the operation of a wooden handpress: she inked the *form* and pulled the *bar* to make *impression;* or she laid the sheets in place and removed them after inking and impression were completed.

There are no data on how Ann Rügerin became involved with her enterprise but there is evidence she may have had male assistance. The central question is: how could she have learned anything about printing independently in a time when such a business was usually conducted as a household enterprise? The probability is that she had lived at some time in a place which had printing within its confines. She did not start her business from scratch, but acquired both her knowledge and her equipment through some other means.

Charlotte Guillard presents an interesting contrast to the foregoing women. She was the ". . . wife and widow of two distinguished French printers and in the year 1552 testified that she had labored in the profession for fifty years."[7] Not having been widowed that long by either of her marriages, the evidence is strong that she had performed work other than that associated with a master-printer, which she became only in widowhood. The title page of *La Vie de Monseigneur Saint Hierosme,* Paris, 1541, says that the book was "Avec privilege, Imprime a Paris au Soleil dor, rue S. Jacques, par Charlotte Guillard, versue de feu [widow of] Claude Chevallon."[8] Another early book, published in 1576, says that it had been printed "In officina typographica Katharinae Theodorici Gerlachii relictae [widow] viduae, et haeredum Johannis Montani," which again indicates how a woman acquired a business, but nothing about any previous printing-office experience except by inference.[9]

An example of succession with specific information is Tace Sowle who started to work as a compositor in Great Britain at an early age. "She succeeded her father, who died in 1693. [John] Dunton says: 'She is both a Printer as well as a Bookseller, and the Daughter of one; and understands her trade very well, being a good Compositor herself... .' "[10] Through the centuries, many other women have successfully carried on printing-offices after the deaths of others, two of the most prominent being the widows of John Baskerville and Giambattista Bodoni."... [W]ives, in the earlier years, though they may have mastered the technicalities and financial details of their husbands' establishments, usually became conspicuous only after they were widows."[11]

The records of the long-lived printing-office established by Christopher Plantin, in 1555, generally afford more of a negative than a positive source of data about women in printing. Leon Voet, in his massive study of the firm, says that "Worthy of note is the fact that in 1583 to 1584 the collator [one who folds and arranges printed sheets] Merton Gilles was assisted by his daughter. This is the only instance ... of a woman in the essentially masculine world of the Plantinian printing shop during the three centuries of its operation."[12] A missing word here is "employee," because earlier in his study Voet discusses the Plantin daughters, and especially the fourth daughter, Magdalena. In 1570, at thirteen years of age, her father said she "... still keeps to the rule that the others have kept to until the same age: that is to say, helping her mother with the housework and principally with her own special task of carrying all the proofs of the great Royal Bibles to the house of Monsgr. Doctor B. Arias Montanus and reading from the originals in Hebrew, Chaldean, Syriac, Greek and Latin, the contents of said proofs, while Monsignor diligently observes whether our sheets are in a fit state to be printed. And the said Royal Bibles being completed by the Grace of God, I intend (from such time as her age no longer permits me to leave her in the company of the proof-readers[13]) to employ her in helping and assisting me in looking after the work that is being printed here and in paying the workmen their weekly wages on Saturdays, and in seeing that every member of the firm does the task expected of him." However, Voet cautions that "All [Plantin] said was that his daughters read proofs, not that they understood them... . The girls' reading and correcting of proofs thus consisted simply of comparing characters, and implied no understanding of the texts."[14] This comment overlooks Plantin's specific description of Magdalena's work as copyholder, the one who reads copy to another who does the actual checking and marking of the proof.[15]

The Ms. Gilles who worked as a collator in the Plantin printing-office was a forerunner in a trade which eventually became dominated by women. For hundreds of years, the finished sheets were hung to dry, collated, stored, shipped or sold individually, thus affording a variety of tasks which were well within the capabilities of women and children. With the advent of edition binding in the early 19th Century, a whole new industry was established, often as an adjunct of a printing-office but, as time went on, more frequently as a separate enterprise. The employee rosters were dominated by "bindery girls," a sexist designation which is commonly used to this day, regardless of age or marital status. The Gilles daughter may have been atypical because she came from outside the family. However, it is possible that when there were no women or children available to assist in a printing-office, others may have been engaged as supplemental labor.

Printing in the New World originated from a contract of 12 June 1539 in which Juan Pablos, an employee of the House of Cromberger in Seville, Spain, agreed "to maintain in the ...

City of Mexico a house and press in order to print books... ."
Of special interest is a paragraph which follows: "Item that the
said Gerónima Gutiérrez my wife is obligated to supervise and
serve in the house in *everything that may be necessary* [emphasis
supplied] and without receiving for this wages or anything
only save her maintenance[.]"[16] This noteworthy contract formalized a business relationship between a master-printer and
his wife with many implications, not the least being the kind
of work she actually performed for the printing-office once it
was established.

Two centuries later, Benjamin Franklin wrote that "[Deb-
orah, his wife,] assisted me chearfully in my Business, folding
and stitching Pamphlets, tending shop, purchasing old linen
rags for the Paper-makers, &c. &c."[17] This description enumer-
ates the kinds of jobs that were probably done every day by
women and children during the many years when printing
was a household business. Elsewhere in his writings, the ever-
practical Franklin reiterated his belief that women should be
trained in business matters for it would "... be of more Use to
them and their children in Case of Widowhood than either
Music or Dancing ... ."[18]

His brother James's wife, Anne, is of special interest because she "... was aided in her printing by her two daughters ...
[who] were correct and quick compositors at case; and were in-
structed by their father whom they assisted. A gentleman ... informed me that he had often seen her daughters at work in the
printing house, and that they were sensible and amiable wom-
en."[19] This is the earliest eyewitness account of American wom-
en setting type, which became a widespread occupation for
them during the 19th Century. According to Lawrence C.
Wroth, "There are to be found on record other instances of
women compositors in sufficient number to make it certain
that in reckoning the labor resources of the colonial printer
the women of the family should be counted on as a possibility
he was not likely to overlook. At this time, indeed, the printing

trade was still in the household stage of development, and it was not remarkable that the women of the family and the well-grown children should be called upon for assistance in the routine of the shop."[20]

Another historian noted that "Although at the period of the American Revolution it was not customary to employ females in printing-offices, yet a woman 'master-printer' was not an uncommon thing."[21] A comprehensive survey of the period 1639–1820 found that twenty-one of the twenty-five women studied ". . . took up the reins of the business relinquished by a deceased husband . . .", thus following the centuries-old practice.[22] The study also found a lack of personal data on American women printers of the period and notes well that "Fortunately, at some point in their lives they learned from some member of their family the rudiments of typography and/or managing a [print] shop."[23]

With the advent of the 19th Century, two classes of printing firms appeared which were distinct in their organization, machinery, hours of labor and wages: newspaper and periodical establishments and the book and job offices. Printing no longer being a household undertaking, women sought work elsewhere in the burgeoning trade. They met strong resistance from the very beginning, especially from unions, because many men saw them as a source of cheap labor and therefore a threat to their own wages and employment. Others felt simply that woman's place was in the home. Women found most of their work as typesetters in the book and job establishments where wages were consistently lower than on the newspapers. Ava Baron, in her Ph.D. thesis, observed that the ". . . development in the printing industry between 1830s and 1850s led to the virtual exclusion of women printers from metropolitan daily newspapers, with women printers confined to the country newspaper and to the book and job branches of the industry."[24] No longer able to acquire either knowledge or skills by daily association in the home, women

had difficulty in learning printing and thus the problem of apprenticeship arose and remained very contentious. By 1870, some locals of the Typographical Union admitted women to full membership and stumped for "equal pay for equal work." There was no altruism in these moves because women were inevitably gaining in numbers in the work force while the equal-pay stance was principally a means to protect the men's wage scale. The International Typographical Union finally removed any ban on women in 1883.

Closely tied to all these concepts was the number of apprentices per journeyman in a given shop, another method of keeping the wage structure steady. Apprentices were always paid on a lower scale, and sometimes earned nothing in non-union shops while learning to set type. Limiting the number of apprentices established a form of control, so that no firm could consistently underbid others because it had so much cheap labor.

Isaiah Thomas wrote that as early as 1815 there were employed ". . . in a printing house near Philadelphia, two women at the press, who could perform their week's work with as much fidelity as most of the journeymen."[25] Soon after, others were reporting women "superintending the printing of sheets by a press worked by horse power" and five years later one of the women lost part of a hand when it became entangled in printing machinery.[26] After the invention of the power press, many women became *press feeders* — those who position the sheets for printing — a simple, repetitive task which was easily handled by young girls and women, at low wages. The traditional handpress had offered a similar job opportunity when

Fig. 3, facing page: *This photograph may be unique in the annals of women in printing in the 19th Century. All are in their Sunday best, and the woman typesetters hold their composing sticks in their gloved hands. The original is a photograph tipped into the "Carrier's Address" for 1887 of the* Santa Barbara Daily Independent. *The pressmen are shown with their locked forms, and the newsboys, editor, bookkeeper, and resident poet round out the staff.* Author's Collection, courtesy Clifton F. Smith, Santa Barbara.

operated by two persons, one of whom put the fresh sheets in place and removed them after printing, a simple task, but one requiring some height, thus eliminating smaller women and children from the work. Pulling the press, impression after impression, was vigorous and tiring work and not recorded reliably as being done by women on a sustained basis.

A Boston newspaper estimated that by 1831 a total of 200 women were employed in printing in that city without a detailed breakdown of the tasks categorized as "printing." Of the 569 compositors at work in Boston in 1849, there were 88 women, 22 of whom worked on general book and job work while 66 worked on weeklies, semi-weeklies and other periodicals, all of which paid a lot less than newspapers for both men and women.[27]

In 1910, Helen C. Sumner prepared a history of working women for the Bureau of Labor. She said: "By 1864, partly, without doubt, as a result of the Civil War, the introduction of women printers began to attract considerable attention. Three other causes of the employment of women were, however, prominent. The first and most conspicuous was . . . using them as strike breakers. The second and probably the most important was the fact that women would do the same work as men for lower wages. The third was the influence of the newly invented typesetting machine."[28] By this time, too, women's specific position in the trade had become established, for they rarely learned anything of the work of the compositor except setting *straight matter*. This is continuous text as opposed to *display matter*, title pages being a familiar example. As a rule, women did not *make up* or *impose* forms for printing; that is, they did not make lines of type into pages or lock them in a *chase* — a rectangular metal frame — in proper sequence for printing.[29] This meant, of course, women generally could not work the other jobs which almost always paid higher wages. A survey of Southward's (British)[30] and Ringwalt's (American)[31] 19th Century printing dictionaries shows more than two-dozen occupations within printing-offices of that era — *excluding all binding operations* — which could be, and often were, included under the generic term, printer. Likewise, the term compositor also covered many different and distinct tasks in printing. By the time of complete industrialization, women specifically worked mainly as typesetters and were always a small portion of the total labor in the printing trade.

Fig. 4, next page: *A period piece in a wide variety of types, the hallmark of much 19th Century printing. Printed by Women's Union Print.* Courtesy The Bancroft Library.

LOWEST RATES
TO
SACRAMENTO
AND
STOCKTON!

QUICK TIME!
COMFORT,
Economy and Speed.
Save your Money!
No Mud!
NO DUST!

FOR SACRAMENTO,
THE NEW AND ELEGANT STEAMERS

YOSEMITE | CHRYSOPOLIS
Capt. E. A. POOLE, | **Capt. A. FOSTER.**

FOR STOCKTON,

AMADOR | JULIA
Capt. CHAS. THORNE. | **Capt. WM. P. BROMLEY.**

Leave Broadway Wharf, Daily,
AT 4 O'CLOCK, P. M.

Calling at all intermediate points; and connecting with all the early Trains, Boats and Stages.

STATE ROOMS AT REDUCED RATES.
FARE.

Cabin,	$1.50	State Rooms,	$1.00
Deck,	$1.00	Single Berths,	50

Women's Union Print, 424 Montgomery St.

Chapter 2   The Many Facets of Printing

The Song of the Printer
Pick and click
Goes the type in the stick,
As the printer stands at his case;
His eyes glance quick, and his fingers pick
The type at a rapid pace;
And one by one as the letters go,
Words are piled up steady and slow... .[1]

This 1870 poem gives only a small idea of typesetting, the principal occupation of the few women in the trade at the time, and nothing of the fundamentals upon which the work was based. In order to understand the setting of type, a comprehension of the basic concepts of letterpress printing from movable types is necessary. First, it is a mirror-image process so that everything must be prepared in reverse. Second, *all* materials must be rectangular and plane-parallel in *all* dimensions. Third, everything must be made to a mathematical accuracy. Fourth, the elements to be printed must be locked temporarily into a solid state — *form* — to sustain impact — *impression*. Fifth, creating a solid state and dismantling it must be relatively simple and quick tasks. Everything else follows from these principles.

Once Gutenberg conceived of printing with movable types in the 1440s, he not only had to develop all the equipment but also to overcome one basic problem: the nature of a Western language in which many letters of the alphabet varied greatly in their lateral dimension — *set* — "m" and "i" being examples. Furthermore, typesetting required variable spacing between words in order for the lines on a page to be the same length — *justified*. Gutenberg's solution was the adjustable-width type mold. It *always* cast types and spaces uniformly in two dimensions but easily varied the third to accommodate alphabetic differences as well as to create the varying sizes of spaces needed to make typesetting mathematically feasible.

The next problem was the metal for type. The formula had to meet the rigid requirements of printing procedures, yet be practical and economical. First, the metal must be reduced to a liquid state at a relatively low temperature; second, it must return to a solid state quickly; third, it must cool uniformly in order not to change *any* dimension inconsistently. If type expanded one-thousandth of an inch in height in one instance or shrank an equal amount in another, the first might print too boldly and the second might print too lightly. Lead, the principal ingredient, tin and antimony, with traces of iron or copper — depending on the era — became the standard metals for type very early. Lead was inexpensive, readily available and melted at a low temperature (compared to brass, for example). Tin regulated the flow of the molten metal to assure a clean, sharp image. Antimony controlled the expansion/shrinking problem, while the iron or copper added hardness. Most importantly, all authorities now agree that the adjustable-width mold and the metal formula for it were the heart of the invention of printing with movable types. All else was accessory.[2] Ink probably was an early problem for Gutenberg because writing inks were always thin and watery and would not adhere to metal types. Artists in the Germany and Low Countries of his time had the finest of paints and, hence, it is not surprising that modern analysis by cyclotron shows a large concentration of copper (oxide) and lead ("white lead") in the ink used by Gutenberg for his *Bible*, definite clues to its origin.[3]

Except for designs, type production remained basically unaltered until 1840 when the Bruce typecasting machine, an American invention, met the rapidly growing need for types in an era of increased literacy and the proliferation of books, magazines and newspapers. Bruce's comparatively simple machine speeded up the casting process by harnessing the conventional mold to a hand-cranked device which pumped the molten metal uniformly. Significant mechanical typesetting only arrived subsequent to 1884, with the invention of the *Linotype*, more than 400 years after Gutenberg.

The printing press was a comparatively uncomplicated adaptation. The principle of a screw mechanism to apply the minimum pressure (75 pounds-per-square-inch) needed for letterpress printing was long and widely known, from the olive and wine presses of antiquity to the papermaker's and the bookbinder's presses of later eras. The problem was to adapt this technology to a printing device which would allow ease and consistency of repetitive operation. Gutenberg brought his concepts together so successfully that from the appearance of his 42-line *Bible* in 1455, to 1500, no less than ten million copies of 40,000 different books had been printed in 260 places in Europe.[4] All this accomplishment was possible because a prime necessity, paper, was now readily available in many cities.

Some scholars believe that Gutenberg's original press was a "one-pull" device, that is, a single pull of the *bar*—handle—printed the whole form. Most agree that his folio *Bible* was printed one page at a time and thus each sheet went through the press four times, meaning the work was done in a one-pull procedure. About thirty years later, the two-pull press became the norm when handpresses were fitted with a smaller plane-parallel surface — the *platen* — which could print twice its dimension in two separate *pulls*, the *bed* bearing the form of type being positioned easily between pulls. Both efficiency and quality improved. This was the only *basic* modification in the wooden handpress during its long history. (The first printing press in California was a wooden, American-made Ramage, 1839.) In 1800, Charles Mahon, Third Earl of Stanhope, invented a stronger iron handpress with a compound-screw mechanism that furnished far more power for the same amount of effort by a pressman. American makers soon followed with their own distinct versions and the one-pull configuration again became standard. The era of the iron handpress was short-lived because of the rapid development of mechanization.[5]

The mechanical *cylinder-press* — where the paper and type touch only at the tangent point — was perfected by Friedrich Koenig, 1810–1812, and had its initial success in 1814. Together with the creation of treadle- and power-operated, platen *jobbing presses* by George P. Gordon and others around 1850, the vocation of *press feeder* gradually expanded and master-printers often trained women and girls for this work.

Typesetting was always a co-operative endeavor because books, newspapers and job work all required a number of laborers whose simultaneous, collective efforts created the masses of type required to keep pressmen busy. In 1586, Christopher Plantin had three presses, six pressmen and four compositors, with two workers always assigned to each press. This ratio of compositors to pressmen was probably fairly constant during the handpress years.[6] The advent of typesetting machines at the end of the 19th Century finally enabled a single worker to set type ahead of the pressmen, depending on the job.

Typesetters normally worked standing in front of a frame or stand which contained several cases of type. The top held the two cases in use: the *lowercase* at a shallow angle, six inches higher at the back, and the *uppercase* at a much steeper angle to the rear and slightly above the other. The uppercase, which contained the capitals, small capitals and special characters, could be farther away because its contents were not as frequently used as those of the lowercase.

The compositor's case consists of a shallow tray two-and-a-half feet long, and half as broad, divided into compartments for the different characters used. *Harper's Monthly*, in an 1865 article describing their printing operation, gave excellent detail: "The lower case, as arranged for ordinary work in English, has 54 boxes of different sizes; these contain the various small letters (hence styled 'lower case letters'), the marks of punctuation, the figures and spaces...of different sizes. The upper case has 98 boxes of uniform size. These contain the capitals, and various [special] characters which are in frequent use, such as parentheses, reference signs, dashes, dollar and pound marks, and so on, while leaving a few boxes for characters which may be frequently wanted for special work. A pair of cases laid [with types] for usual work contains about 140 *sorts*."[7]

The alphabetic characters in the uppercase were in strict order except for J and U, which followed Z, a perpetual reminder that these were the last two letters to join the English alphabet.[8] The lowercase, after many years of changes to attain maximum efficiency, finally became standardized in the 19th Century, and the boxes for letters and spaces were sized and grouped according to frequency of use.

The standard *job case* and the *italic case*, used for special printing, combined capitals — but no small capitals — lowercase and a few special characters. Typesetters often put a job case in the lowercase position temporarily. An italic case usually was present in the top part of the compositor's stand to afford quick accessibility for the occasional word which might be needed in italic when setting straight matter.

Each piece of type was cast with a *nick* in its lower shank which designated the bottom of the character. More than one nick indicated a different *font*, such as italic. Both by glance and feel the typesetter would know that all the characters were right-side-up in the stick and that there were no non-matching characters — *wrong fonts.*

In 1871, Luther Ringwalt described handsetting procedures: "First, let the letter on which the eye first rests be picked up and secured in the fingers, before the eye moves to another box. Second, let the eye be fixed on the second letter while the right hand is bringing home the first; and fix it on the third while conveying the second, and so on. Again, let the time of picking up a letter be equal to the time of counting two, and bringing it to the stick equal to one, counted at the same rate. ...

"The general directions are to take up the letter at the end where the face is; if the nick is not upward, to turn it upward in its progress to the composing-stick; and to convey it to the line in the composing-stick with as few motions as possible. ... The fast compositor, for instance, is more apt to drop his type in his stick [hence the "click"] than to laboriously place it there; and while his right hand is picking up the type [the] left is advancing...with the stick to receive it. ..." Thus the typesetter lost no motion, though viewing the types upside-down in the stick. For maximum efficiency, the stick always was moved toward the right hand to shorten the distance from box to composing stick. Ringwalt concluded his picture of the working compositor: "Standing perfectly upright before his frame, soldier-like, heels and knee-joints together; the upper part of the body erect, not curved; looking at a type before [the] right hand and finger grasp it, and then bringing it to the composing-stick ... ."[9]

As Ringwalt implies, the posture and position of a compositor were most important. Another authority added that "the height of the compositor and his frame should be so adjusted, that his right elbow may just clear the front of the lower-case by the 'a' and 'r' boxes [immediately to the right of center and nearest to the worker's right hand], without the smallest elevation of the shoulder-point…[while another master-printer] states that personal experience . . . has fully satisfied him that at a frame the height of the elbow, a compositor will do his work quicker, easier and hold out longer than at a high one."[10] Thomas Lynch (1859) gave a precise detail lacking elsewhere: "The position, when setting types, should be such that the right side of the compositor would be in a line with the central division of the lower case, because he will be able to reach the boxes at the left-hand side with more ease… ."[11]

Thomas MacKellar had his own specifics in the 1870 edition of his manual: "The standing position of a compositor should be perfectly upright, without stiffness or restraint; the shoulders thrown back, the feet firm on the floor, heels nearly closed, and toes turned out to form an angle of about forty-five degrees. Habit will render a standing position familiar and easy; perseverance in conquering a little fatigue will be amply repaid by the prevention of knock knees, round shoulders, and obstructed circulation of the blood… ."[12] An earlier British writer noted that "However zealous, therefore, a workman may be, if his shoulders and hips are seen to be moved by every little letter he lifts, fatigue, exhaustion and errors are the result."[13] The health and production concerns were certainly important considerations in an era when the work schedule was a ten-hour day, six days a week.

Another manual summarized the positioning of the typesetter succinctly: "The caps should be breast high; this prevents rounding of shoulders (the general mark of compositors) and also complaints which result from pressure by bending the breast. The standing position should be easy — feet not too much apart — nor should either foot be rested on the frame, or turned inwards; a habit of doing either insures deformity of the limbs. The head and body should be kept perfectly steady, the arms alone performing the labor."[14] That typesetting was serious business is reflected in a final caveat from another 19th Century manual which declared, "All singing, whistling, or unnecessary conversation, during working hours, peremptorily prohibited."[15] Another bad practice should have drawn comment, but apparently did not. Unconcerned with the hazard of lead, apprentices occasionally held types and spaces with their lips during *justification*.[16] The master-printer, foreman or forewoman supervising the training should have routinely forbidden this practice.

All who set type confront the basic fact that printing is a mirror-image transfer and therefore type, already cast with a reverse image, must be set from left to right with the first line at the bottom. To meet these criteria, type is always set in the simply conceived *composing stick*, which is a rectangular-shaped, shallow tray with sides enclosed and the topmost, long side, open.[17] It was designed to be grasped completely in the left hand with the thumb free to hold individual types in place as the right hand brought them into the stick. (Left-handed persons had no choice but to conform.) The lateral dimension — the *measure* — was always constant for a given project.

A typesetter held the stick efficiently when its top edge was higher than the portion next to the palm of the hand and the whole slightly tilted to the left. Experienced typesetters gradually increased the vertical angle slightly for balance and to take the strain off the wrist as the stick became heavier with lines of type . Most importantly, holding the stick at an angle took advantage of gravity. Consequently, a single type — often very small — stayed upright more easily when released from the fingers of the right hand until the left thumb could apply pressure to keep it firmly in place. The basic construction provided room for comparatively few lines at a time, depending on the size of the type being set. This made the composing stick easier to grasp, and diminished the fatigue generated by holding undue weight over the standard ten-hour work shift.

The spacing between words was based on the *em*. By definition, this was the square of the body of a given type size. An em can be divided into several equal portions, mathematically related, and so typefounders were able to make the different widths of spaces in an orderly manner: 3-em — short for 3-to-the-em — 4-em, 5-em and 6-em spaces being common to all sizes of type. The *en quad* — short for *en quadrat* — was half an em and 2- and 3-em quadrats were multiples of the basic em quadrat, and useful — less pieces to pick up — in filling out the ends of paragraphs, poetry lines and the like.

The correct use of differently sized spaces to justify each line exactly required the typesetter to learn the mathematical relationships of spaces.[18] "A line of type properly justified will stand up in the stick at any angle without falling over and still will be easy to remove without pieing [scrambling]," according to a Union training manual.[19] With line after line set in this manner, the mathematical exactitude to lock up the types into a solid state — the *form* — was accomplished. The composing sticks of typesetters — by custom, many brought their own to the job — had to be accurate. They accomplished standardization of measure for many years by moving an adjustable slide (always on the left) to a steel, or sometimes brass, *compositor's rule*.[20] A complete set of these rules — also frequently owned by the worker and carried in a felt-lined, oak case — contained every standard measure for text work from the narrowest to the widest. The technique of proper justification determined the tightness of each line — how difficult it was to push the last character or space into place. Consistency came only from experience and, literally, a feel for the task. Correctly set type is easy to read with no unsightly spacing, such as "rivers of white." A reader never notes the best composition.

Removal of the lines of type from a composing stick also required dexterity, for a stickful might consist of 1,000 or more separate pieces. The typesetter pressed the ends of the type lines with the sides of the second fingers of each hand, at the same time squeezing the top and bottom of the lines (using the thumb and forefinger of each hand) between strips of metal — *slugs*. These actions enabled the typesetter to put the types into a solid state temporarily in order to move them, either by sliding them out of the stick or by lifting them for carrying. It was much easier to handle the types after the adjustable, quickly removable slide was invented in the 19th Century.

The typesetter regularly transferred lines of type from the composing stick to a *galley*, a three-sided, shallow, metal tray, commonly $23^{1}/_{2}$" long and of a width suitable for the book or news work at hand, one short dimension being open.[21] A galley normally would be placed on a *rack* or *bank* set at an angle — another way of using gravity to keep the types and spaces from falling over. Gradually, the accumulating lines filled the galley sequentially, stickfull after stickfull. When full, it was locked and proofed — hence, *galley proof*. The predetermined number of lines, including running heads and page numbers — *folios* — needed to create a page governed *making up*. The typesetter thus had to fit a measured pattern. Other material, such as tables, although set in different measures, had to meet exactly the physical criteria for the project at hand in their totality.

The narrow columns of newspapers presented special problems for typesetters who constantly made more decisions regarding spacing and hyphenation than those setting wider measures. For example, more than two hyphens at the ends of consecutive lines in book work was a sign of a poor typesetter, while three hyphens in a row were acceptable in newspapers. A good typesetter had to have a sound understanding of both the theory of spacing and English grammar to accomplish these never-ending tasks quickly and correctly.

Once galleys of type were proofed and corrected, a journeyman compositor made them up into pages and locked them with wedges — *quoins* — into a rectangular metal frame — *chase*. The resulting form was then tested for *lift*. An object

— commonly a *quoin key* — was placed under one corner of the chase to raise it slightly. If justification was improper, types would fall down and this form could not be safely carried to a press. The compositor further tested the form by pressing all over each page. If there was "give," it meant spaces could *work up* during the press run and print unsightly black marks. In both instances, the offending lines had to be removed to a composing stick for rejustification, replaced in the form and the whole again tested for lift. Improper typesetting thus cost everyone in time and money.

For more than 500 years, printing was a very conservative trade because of the repetitive nature of typesetting and presswork. The early practices quickly became codified into rules which master-printers and their employees understood as a kind of verbal folklore. Joseph Moxon, in 1683, first elaborated them extensively in his still-studied *Mechanick Exercises*.[22] In recent times, Lawrence C. Wroth, a well-known librarian, noted a singular fact: virtually all the principal printers' manuals, from John Smith's *The Printer's Grammar* (England, 1755) to Thomas MacKellar's *The American Printer* (1866 ed.), contain large sections of Moxon which were either taken verbatim (by Smith) or summarized or paraphrased without disguising completely their Moxon origins.[23] Wroth thereby was able to highlight the unchanging nature of many basic printing practices as they spread from 17th Century England to the United States of the 19th Century.

The direct effects of the Industrial Revolution on basic printing technology were confined almost exclusively to presses in the first years of the 19th Century and are reflected in the printers' manuals. It was the typesetting procedures which remained closely tied to the past, even with the advent of the composing machines much later in the century.

The only mentions of women — non-working — by Moxon occur in "The Customs of the Chappel [the group of workers in a given shop]" where an elaborate series of contributions are described, including the following: "If a Journey-man marry, he pays half a Crown to the Chappel; When his Wife comes to the *Chappel*, she pays six Pence; and then the Journey-men joyn their two Pence apiece to welcome her; If a Journey-man have a Son born, he pays one Shilling. If a Daughter born, six Pence."[24] (What price woman!)

Because so many practices were customary, specific procedures for given eras and places are often unknown today. For example, no one thought it was necessary to record every step from the intake of copy to the delivery of the finished work in a specific shop. The precise manner of figuring payment for typesetters in 19th Century America remains difficult to reconstruct, while there are data for Great Britain.[25] The possibilities for unending disagreements were many, especially if there was an exploitative or dishonest employer. The literature of printing is singularly devoid of this kind of problem except for the perpetual quarrels over *fat, lean, dirty and clean copy* (see below) which could affect both the workers' pay and shop efficiency. Ringwalt says these arguments often were resolved by the ancient practice of *jeffing*, which is "throwing with quadrats, somewhat in the fashion in which dice are thrown, to decide disputed points in printing offices, such as who shall receive a fat take . . . ."[26]

A typesetter's rate of pay was traditionally based on 1,000 ems. Clearly it was easier to earn more with copy which had many blank spaces, such as title pages and poetry: hence, *fat copy*. The counterpart was primarily solid text: hence, *lean copy*. The Typographical Union spent years arguing over the differences in setting various sizes and styles of type based on the measurement of the lowercase alphabet, which differs markedly, in the same size, from design to design.

Dirty copy was a constant problem. If a typesetter could not easily read a hastily written manuscript or a poorly trained hand, the extra time required to decipher it led to a loss of productivity. The worker normally lifted the next piece of copy —

the *take* — from a curved spindle in the composing room, the *hook*, where it hung after editing and marking up for size and style of type. This random system developed to treat all equally, but much mischief could result from someone having changed the order of the copy on the hook. If a typesetter made such a move, it was called " 'sojering' or finessing for a choice of *takes*."[27] Although every typesetter "is supposed to 'follow copy' rigidly, he does not and cannot always do so," a Pacific Coast trade journal declared. "[He] really assists at editing . . . . The compositor is expected to be a walking encyclopedia, a philological prodigy and a decipherer of genuine hieroglyphics."[28]

No one could become a fast typesetter who could not readily follow the copy. For efficiency, the trick was not only to remember the text in small segments, but also for the eyes to return quickly to the exact point from which a new segment would be memorized. Although a number of patented copyholding devices were available in the 19th Century, the most common practice was to prop the copy against the small-capitals portion of the upper case and to use a long slug to hold it in place.[29] This part of the process was highly individualistic. The nature of the whole procedure makes Benjamin Franklin's great invention, bi-focal glasses, a boon to older typesetters.

The rate of pay traditionally covered the *distribution* of the type into the cases after use. Journeymen were frequently as inconsistent in this chore as apprentices, and their carelessness led to characters becoming mixed up, resulting in a *foul case*. Certain letters, such as *b, d, p, q, n, u*, and punctuation, such as ' (apostrophe) and , (comma), were perennial problems because of careless distribution. Spaces, too, were often improperly distributed; it took an experienced eye to differentiate the smaller ones correctly with a quick glance. A typesetter working from a foul case inevitably was slowed down by non-payable corrections. Experienced typesetters preferred to work at their "own" cases.

Although typesetting remained piece-work for a long time, the procedures used in San Francisco to determine the number of ems set and the bookkeeping involved are not known. The extensive records of Towne & Bacon, at Stanford University, show only a daily rate of pay for all employees, from 1857 to 1875, with no hint of the 1,000-em method of paying for typesetting.

Under the heading "What Can Girls Do?" the *Woman's Pacific Coast Journal* gave a succinct summary of printing realities in San Francisco in 1871, when women had become more visible in printing-offices: "To be a printer, a person should be a good grammarian and be accurate in orthography and the rules of punctuation. A person entering an office at the age of eighteen or over will be required to give their time for three months; they then can earn their board for three months, after which they expect small wages. With care and diligence, they will become in a year's time good compositors for newspaper work. A job printer requires correct taste and longer practice.

"Newspaper work is done by price in this city, at fifty cents per thousand ems. From five to eight thousand ems may be considered good work for a day of ten hours. Job printers work by the week, at from twenty to thirty dollars."[30]

This description probably reflects long-standing pay practices in San Francisco and thus could explain why the Towne & Bacon records show *all* workers being paid by the day. Further, payment by the day allowed an employer to make a markup on the composition by billing it to customers as ems set.[31] The strong hold of the Union on newspapers, as well as the repetitive nature of the work, undoubtedly served to perpetuate the payment-by-ems tradition at these offices. There is also evidence that when a book-and-job shop had large amounts of type to set for a given book or publication, management used the "by price" method.[32]

The typesetter, being responsible for making galley

proofs of all of his or her work, pasted a duplicate set — the *dupes* — of the corrected proofs into a continuous strip — the *string* — which the Providence, Rhode Island, Union local defined as "the pasted and measured result of a printer's day's labor . . . ."[33] A San Francisco typesetter, Maggie Welsh (A 276), testified in 1888 that "I worked three weeks for Bacon [& Co.], receiving 30 cents per 1,000 ems. I measured my own work...."[34] On the other hand, the Typographical Union, in a set of printed grievances against Bacon & Co. in 1888, said that Bacon had been "refusing to allow the girls the privilege of measuring the matter they have set."[35] The use of the word "privilege" is especially noteworthy in the context of Bacon's low wages.

Other mentions by males or females who gave testimony at the labor hearing did not indicate a basic pattern, nor do the standard printing reference books of the 19th Century. Apparently, everyone understood how it was done — whatever the variation in the procedure from shop to shop — and had no complaints. In small firms, a typesetter probably measured his or her string; in a large printing-office, the string most likely went directly to the person in the front office responsible for payrolls. Strings were usually measured with a "triangular, six-sided ruler similar to those used by draftsmen [with two scales on each side] except that scales were in *picas* — 12 points — and other type sizes commonly used in text work."[36]

Ringwalt notes that "Various tables have been published, having for their object the saving of time and labor in measurement, and obtaining ready results in calculating differences in proportion of the various sizes of types."[37] Lynch's manual offers two convenient tables: the first "showing the number of lines contained in 1,000 [pica (12-pt.)-ems] in all measures from 10 [pica-ems] to 100 [pica-ems] wide;" the second "showing the number of [pica-ems], in the various kinds of types, which will occupy the same space as 1000 [pica-ems], in all the sizes from pica to *nonpareil*[6-pt.]."[38]

The basic unit in printing is the *point*, an arbitrary entity finally standardized, in 1886, by most of the type foundries in the United States, at .013837", approximately $^1/_{72}$". All sizes are expressed in points and thus 12-pt. type would have a 12-pt. em, which means its width and depth are both 12 points. The point was crucial to the method of reckoning the amount of type set. The Typographical Union complained that "The number of lines is an unreliable unit of measurement, because lines are of various lengths, of various sizes of type, and of varying degrees of difficulty in spacing and justifying. Lines in extremely narrow measure, for instance, are much more difficult to set than lines of greater length. To overcome these varying factors of type size, face, and length of line, printers have adopted the em as a standard unit for measuring the surface, or area, of type composition.[39] To figure ems of solid matter, find the number of ems in one line of the type size being used and multiply by the total number of lines in the composition. The ems in one line may be found by multiplying the measure [length of line delineated in picas — 12 pts. = 1 pica] by 12 to get the total width in points, and then dividing by the size of type to get the number of ems in one line. ... Fractions of an em in the length of line are counted as full ems."[40]

On the other hand, the noted printer-scholar, Theodore Low De Vinne, declared the em was ". . . a unit of superficial measure. The space that can be covered by one thousand em-quadrats is reckoned as one thousand ems. This method of measuring is never changed for open, or leaded composition [with spacing between the lines of type — fat]. One thousand ems may contain three thousand bits of metal if the matter be solid [no space between type lines — lean], or only one thousand bits if the matter be leaded and full of quadrats [larger spaces]; but in either case the composition is counted as one thousand ems."[41] After a lengthy discussion of how the differences between type styles affected measurement, De Vinne concluded: "The term one thousand ems ... does not fairly describe the amount of a compositor's labor, or even approximately the number of words in his composition. Under present standards the compositor of books has to set from one fifth to one half more matter than the news compositor to have it rated as one thousand ems."[42] This fact had an important bearing on work opportunities and take-home pay for all typesetters in San Francisco and elsewhere in the 19th Century.

De Vinne also wrote specific instructions for the measurement of pages: "All pages should be measured from extreme points. . . . A rule-bordered page should be measured from the outer points of the rule; pages in columns should include all spaces between columns; chapter heads or tails, or any blank space, or [wood-]cut, within the area of the page, should be related as type."[43]

Starting around 1830, compositors in book offices gradually lost some fat when workers designated as "printers," and paid by time, took over making up of pages and completing forms — *imposing*.[44] Consequently, the compositors in such shops became typesetters only, work that required less training, but which offered higher pay than other industrial employment available to women and was well within their capabilities. Also, women potentially represented cheaper labor in a high-cost market, such as San Francisco. However, until 1870, the Union would not tolerate their working with Union men, even at Union wages, while the same prejudice also kept females out of numberless non-Union firms. As a result, most women remained constantly on the fringes in both jobs and wages until the 1880s. However, commencing in 1868, woman-run printing-offices in San Francisco consistently afforded a significant number of opportunities for female typesetters, including the all-important training.

Printing is a manufacturing business and therefore productivity has always been a primary concern. The authorities and manuals cited above on typesetting represent the distilled experiences of many years. There is a large common denominator, but practices probably varied widely from shop to shop and there is no evidence concerning how many of the detailed tenets were incorporated into the teaching of woman typesetters in San Francisco. A 90-day training period appears to have been the norm, and how elaborate the indoctrination might have been may well have depended on the immediate needs of a given shop for producing revenue work by using all hands.

Miss Kittle Burke.

Figs. 5 & 6: *From San Francisco* Call. Courtesy The Bancroft Library.

The scant business records that do exist, such as those of Towne & Bacon, show that journeymen woman typesetters generally earned the same amounts as men for a given period of labor and therefore may have been as well trained — or experienced — as their male colleagues for the work at hand.

Miss Kitty Burke (A 36), of San Francisco, became such a noteworthy typesetter that the *Morning Call* devoted a full column to her exploits. The story contains so many relevant details concerning women, typesetting and printing practice in the San Francisco of the latter part of the 19th Century, it is included here in its entirety.

"In typesetting it's Miss Kitty Burke. There's a best in everything, says a wide-awake advertiser. He's right to a dot, and in the art of typographical composiiton by a woman it's Miss Burke. She sets 1,200 to 1,400 ems solid printed matter [8-point] without the leads between the lines, in a single hour. That's not a wonderful record for a male expert, but there's no woman in San Francisco can beat it, say the printers. And they ought to know.

"But it's not the speed alone this good-looking young

Miss Burke at the Case.

lady is famous in her art for, but speed coupled with accuracy. 'Many times we print her composition without ever reading a line of it in the proof,' says C. W. Nevin, her employer. Now, accuracy is a great thing in typographical composition. Two thousand ems an hour can be set so 'dirty' as to take three hours to correct the matter after the proof is read. 'Dirty' proofs are the bane of printers, and the man or woman who can 'set a clean proof' is the fastest compositor ninety-nine times out of a hundred.

"But just the mechanical composition of 1,200 ems of solid matter in a single hour is an achievement away above the ability of the average printer. Your good journeyman printer sets 700 to 900 ems an hour, and if, perchance, he sets 1,200 in the same time he is likely to sweat. The printers in America, men or women, who can set a galley of matter that needs no correction are about as scarce as flies at the north pole. Yet they do exist. Here and there is one, but they are usually slow. Where speed and accuracy are combined there you'll find — well — say, Miss Kitty Burke.

"Miss Burke has been employed by C. W. Nevin & Co., on Commercial Street, near Sacramento, for the past six years. She is yet a young woman, not 23 at most, and gives promise of eclipsing some of the star male compositors for speed, let alone accuracy. 'She's better than nine-tenths of the men printers I've met,' says Mr. Nevin to *The Call* reporter yesterday, 'just as an all-around compositor. Many of the best compositors on plain matter are no jobbers to speak of; they lack the artistic instinct,' continued Kitty Burke's employer. 'The good jobber must be an artist, must have a fine perception of proportion and an eye for the beautiful in color and form. Miss Burke is as good a job compositor as I have ever met.'[45] "Not only is Miss Burke a fast compositor, but a general favorite in the composing-room. Some women are hard to please in any sphere of life, just the same as some men. Miss Burke is not of this number. Her foreman vouches for the fact that she has never yet made a 'kick' on the office towel. Since the time of Benj. Franklin's apprenticeship at Stationer's Hall [London] the office towel has been a legitimate standing joke in all well-regulated printing offices.[46] Mr. Nevin's' office is no exception. Nor is his towel an exception. It stands alone in the corner and is older than his oldest press. That Miss Burke makes no complaint at the antiqueness of that towel, but quietly dries her hands on some scrap paper, proves at once her foreman's claim to Miss Burke's demureness and modesty.

"Miss Burke is a 'clean distributor.' That's a point you probably know nothing about, if you're not a printer. But it's a most important one. After the type has been 'set' it must be 'thrown in' again, or distributed into the respective boxes of the case. Here's where most of the fast printers fail. They distribute fast, sometimes with great speed, but usually with more or less inaccuracy. Carelessly they get an e into the c box, a u into the n box, and so on and so forth. All of which is likely to show up in the proof. Now, Miss Burke is a 'clean distributor.' Without sacrificing speed, she 'throws in her case' carefully and accurately. That's one secret of her speed and accuracy in composition. Therefore, no wrong letters in her type boxes and consequently none show up in the proof.

"If there's a better lady compositor in the city *The Call* would like to know it. The printers say there is none."[47]

Click, click
Go the types in the stick,
They glide in together with an ominous sound,
As quickly the hand that collects them goes round,
And arranges them firm in the stick,
Click, click. . . .[48]

Fig. 7, following page: *Spiritualist title printed by woman-run printing office.* Courtesy The Bancroft Library.

DISCOURSES

GIVEN THROUGH THE MEDIUMSHIP

—of—

MRS. CORA L. V. RICHMOND,

—at—

San Francisco, California, in 1883.

SAN FRANCISCO:

WOMEN'S CO-OPERATIVE PRINTING OFFICE, 424 MONTGOMERY STREET,

1884.

Chapter 3   Woman's Movements and San Francisco

The immediate antecedent for American woman activists was English, and in particular, a book by Mary Wollstonecraft, *A Vindication of the Rights of Women*, 1792, which was soon reprinted in the United States.[1] Several pioneer female reformers immediately appeared to espouse various causes, ranging from the abolition of slavery to the equality of woman in the work place and — later — woman suffrage. Their diverse individual efforts culminated in a convention held at Seneca Falls, New York, on July 19–20, 1848, and growing feminist actions followed. As one historian noted, "In regarding the Seneca Falls Convention as the birth of the movement for women's rights, we are on solid ground only if we remember that birth is a stage in the whole process of growth. In this case the process had begun almost a half a century earlier."[2]

Seneca Falls — about 51 miles west of Syracuse — became the location for the momentous gathering because it was the home of the leading organizer, Elizabeth Cady Stanton, a proponent of woman suffrage. She had decided, with the concurrence and assistance of Lucretia Mott, a Philadelphia abolitionist, that it was time "to advertise feminine wrongs." Sixty-eight women and thirty-two men ended the conclave by signing a long series of resolutions which focused on the idea that "woman was man's equal."[3] Mrs. Stanton herself offered the suffrage resolution: "Resolved: That it is the sacred duty of the women of this country to secure to themselves their sacred right to the elective franchise." It passed by only a slim margin and was the sole resolution not receiving a unanimous vote. One of the 300 attendees was Charlotte Woodward, 19, who aspired to be a typesetter in a print shop and who was the only one present who lived to vote for President of the United States in 1920.[4] The ferment engendered by the historic Seneca Falls meeting endures to the present time.

When Harriet Martineau, a well-known British author and feminist, visited the United States in 1836, she noted that there were only seven occupations open to women: teaching, needlework, keeping boarders, working in cotton mills, typesetting, bookbinding and domestic service.[5] Unlike the work in the cotton mills of New England, typesetting, binding and allied work were jobs women had performed to one degree or another in printing households for more than three hundred years. Women's lives changed drastically when they began to seek employment outside the home in the 19th Century. According to another historian, "The housewife was no longer an indispensable linchpin of society. As the need for her declined, so did her value, and so did the respect accorded her."[6] In common with all women whose economic worth at home diminished with industrialization, those choosing to enter the printing trade found challenges, prejudices and frustrations. Many skills, such as spinning, sewing and domestic chores, had always been strictly regarded as "woman's work" and invariably tasks so classified brought women into the marketplace as cheap labor. The large employment of women in bookbinding, which developed as a separate trade early in the 19th Century, furnishes an example of categorizing special kinds of work as female oriented — folding, sewing — and consequently the pay was low. Nevertheless, "In the long run," said this historian, "the development of industry, despite its initial dislocations and exploitation, would prove a major step in getting women outside the walls of their own homes."[7]

In a study of craft apprentices "from Franklin to the machine age," W. J. Rorabaugh says he "never encountered a female craft apprentice. . . . Poor girls were 'apprenticed' to housewifery or sewing, but that sort of apprenticeship provided only legal guardianship and training in traditional female work rather than the learning of a craft."[8] The limitation of almost all women in printing to typesetting, as opposed to the journeyman compositor's varied tasks, nevertheless put them immediately into rivalry with men.[9] It was an historic first because never before had women attempted to seek and hold industrial jobs in direct competition with the traditional male bread-winners, and consequently there was much friction. The employment of women as typesetters in the United States peaked in the 1890's but they were always a small fraction of the total workers. The percentage of women in printing and publishing — jobs undifferentiated — rose from 9.1% of the work force in 1870 to 11.6% in 1880, 13.9% in 1890 and 17.6% in 1900.[10] Only in the outlying communities, where printing operations sometimes resembled the traditional household style — but often outside a home setting — were women in a significant proportion as typesetters, principally because they usually worked in the printing-offices of newspapers which had very small staffs. Also, they represented cheap, readily available, local labor and were dependable, compared to the traditional tramp printers.[11]

By mid-century, the gradually evolving woman's groups became interested and active in the labor movements. In 1868, Susan B. Anthony, Elizabeth Cady Stanton and Mary MacDonald founded the Workingwoman's Association #1 of New York. The membership consisted principally of typesetters and clerks employed by Ms. Anthony's suffrage newspaper, *The Revolution*.[12] Like many other organizations of the time, its primary interest was suffrage, which put it at cross-purposes to the workers' needs or desires for one simple reason: the reformers were from middle- or upper-class backgrounds and their insights into the difficulties working-class people faced were often distorted or lacking altogether. These advocates saw a large cause before them as the goal while the workers were mainly concerned with more immediate matters, such as work itself and the pay for their labor. Thus suffrage, a long-range objective, was often very low in the female workers' priority of interests, while the suffragists tried diligently to convince them that the ballot box was the salvation for all their problems. In suffrage — and in some labor practices — the West was to be the leader. One historian noted, "The westward-moving frontier . . . continued to provide a freer, healthier and less traditional atmosphere. . . . There was more respect, in the western communities, for women as individuals. When the woman's rights movement finally began to grow, it would be the West that would give it much of its strength and encouragement."[13] In 1869, Wyoming was the first state to extend suffrage to women, followed by Utah a year later, Colorado in 1893 and Idaho in 1896. After long years of advocacy and effort, California finally came into the fold in 1911, by the slim margin of only 3,500 votes. San Francisco males later registered a heavy vote against the woman suffrage amendment to the U. S. Constitution, but in other urban areas they favored enfranchisement.[14] Women generally lagged badly in their support of suffrage right down to the adoption of the Nineteenth Amendment in 1920.

By the 1850's, woman typesetters frequently had gained both training and employment after they were hired as strikebreakers, another long-term source of discord with the Union. During an 1853 strike at the *Day Book*, in New York City, the firm advertised for "girls" to learn typesetting and hired four of the forty who applied. The publisher and editor believed that the wages earned were good and that typesetting was a practical way of enlarging the sphere of female labor. The crux of the matter, however, was the declaration that *Day Book* "would not submit to the tyranny of any union in the universe."[15]

An extreme instance occurred in 1849 during a strike at Bolles & Houghton in Cambridge, Massachusetts. A woman, who "worked at the case to fill up her extra time when not employed in [proof]reading," was asked by the Boston Printers' Union to relinquish her job and the Union offered to pay her board in exchange. She refused and then they threatened her with tar-and-feathering but she still held her ground despite intimidating letters from "Striker." Finally, a rock attack seriously injured her as she was returning home. A newspaper report of the episode termed it a "disgraceful outrage."[16] A later strike at the *New York World* resulted in 100 women being hired as an experiment from which the newspaper concluded that "... the majority of them... never reached an equal skill nor earned as much as [the men]... ." The *World* further claimed that none of them had the strength to work more than seven or eight hours per day, ten hours being the norm. "Clean composition was next to an impossibility. . . and the correction of their 'matter' they shirked [this being unpaid work], if possible, and did badly when it could not be shirked." In conclusion, the *World* generously said that all were "neat and decent in their dress and well-behaved, and that most of them had an ordinary common-school education."[17] These comments may have been written in 1868 but the attitudes represented were the norm for many years.

For a time, the Typographical Union decreed that women who received the same scale for typesetting should join locals of the Union, while those who were paid less should be in all-female locals and so the Woman's Typographical Union #1 was chartered in New York City in 1869. The Union abandoned the authorization of separate female locals in 1873 and in 1875 added a section to its General Laws stating that "Subordinate unions cannot refuse women admission on account of sex," a major victory for feminists.[18] Despite the reduced wages consistently paid non-union female typesetters for skilled work, they were still the elite among all woman industrial workers. Their pay may have been less than males received for the same work, but it exceeded any other labor-oriented occupation open to women. They were also selected for their skills in spelling and grammar, plus dexterity — basic requirements for typesetting — which brought better-educated and faster learners into the printing work force successfully, albeit as a comparatively small minority.

Woman typesetters were relatively late in gaining a foothold in San Francisco which Gunther Barth stated "telescoped the cycle of American growth from wilderness to city into a single generation."[19] At its beginning, San Francisco was fundamentally a Spanish community, Anza's expedition having started a colony in what was an Indian village in 1776.[20] It was not until 1822 that the first Anglo-Saxon settler, William Richardson, stayed on as a deserter from the English whaler, *Orion*, and thereafter San Francisco followed a classic pattern; as historians have found, ". . . a fully developed society evolved, complete with manufacturing establishments, accumulated capital, . . . cooperative institutions, mature governmental practices, and cultural outpourings. The resulting civilization, however, differed from that of the East, modified by the accident of separate evolution and by the unique social environment."[21] Acceleration in San Francisco came in the years following the discovery of gold at Coloma in 1848. By 1849, San Francisco had only 1,000 people but ten years later the number exceeded 50,000.[22] Very rapidly it changed from a small, dusty town to a metropolis with diverse enterprises capable of producing a wide variety of specialized items from railroad locomotives to printing types. Supporting and complementary businesses arose which were common in the Eastern cities of the United States at the time but which appeared almost overnight in San Francisco. A busy waterfront attested to the stream of commerce which gave both financial and practical support to the growing city.

Not the least of the changes was the transplant of ideas, and thus the woman's movements also arrived as more and more females came West to join or seek husbands, to find work, to follow freedom or to find adventure. The developing social movements brought women into printing as workers and entrepreneurs in San Francisco. Except for newspapers, which often included a job shop, and contrary to hundreds of years of printing history, a woman rarely became a master-printer by succeeding to a spouse's printing business in San Francisco from 1857 to 1890 (see A 145). On 12 April 1858, the State Legislature made it legal for a woman to enter business independently of her husband but a woman entrepreneur in the printing business did not appear until ten years later and she had no husband of the moment.[23]

Many people identify woman's movements only with suffrage and the workplace. The record shows that San Francisco rapidly acquired many other causes which were often inextricably linked to all of them through key individuals: Spiritualism, dietary reform, dress modification — anti-corseters and the "Bloomer girls" — Free Love, mind healing, cremation, legal reform and many others. All of these ideas came West with the population influx and one of the most pervasive, from the 1850's to the end of the century, was Spiritualism. A leading scholar defined it as "a new religious movement aimed at proving the immortality of the soul by establishing communication with the spirits of the dead."[24] By coincidence, it also had its origin in 1848 (31 March), in upstate New York. The principal difference between Spiritualism and all other religions was the prominent role it afforded women. It was "the only religious sect in the world" which "recognized the equality of women."[25] It was also unstructured, unlike traditional religions, and "attending a séance was an intensely individual action."[26] Another researcher added, "What must always be kept in mind is that Spiritualism reached its greatest audiences in nineteenth-century America when it adopted a language of plain common sense and spoke to the everyday interests of ordinary people."[27] According to Ann Braude, "Spiritualism became a major — if not *the* major — vehicle for the spread of woman's rights ideas in mid-century America." There was much overlap in the leadership as well, San Francisco included.[28]

There were letters to San Francisco newspaper editors detailing Spiritualist experiences as early as 1852.[29] A Spiritualist manual, published locally the same year, had "theoretical and practical information."[30] The well-known pioneer, Mrs. Eliza W. Farnham, gave the initial public lectures on Spiritualism in the city in 1859. The first Spiritualist publication, *The Family Circle*, appeared the same year, and the movement's visibility was enhanced by a succession of others which came and went through the latter half of the 19th Century, and some of them employed women as typesetters.[31] Spiritualists were often Republicans, who advocated the abolition of slavery. During the Civil War, English Spiritualist Emma Hardinge stumped the state for Abraham Lincoln. "After the the Civil War, many American reform movements drifted from their pre-war individualist origins. Spiritualism was tied to individualism by its religious practice, so it became a haven for aging abolitionists as well as radicals after the war," added Ms. Braude.[32] All of the principal figures in this study, and in the suffrage movement, had some connection with Spiritualism, ranging from actual practice to the printing of books or appearing on a public platform with Spiritualists. Many prominent California women, such as Laura DeForce Gordon (1838–1907), an active suffragist and one of the first two female lawyers in the state, were enterprising in Spiritualist affairs, held séances and lectured widely on the subject.

"Woman suffrage benefited more than any other movement from the self-confidence women gained in Spiritualism."[33] Spiritualism was also inextricably linked to dress reform and one lecturer proclaimed that " 'physical reform' must
accompany spiritual emancipation. . . . Confining clothes kept
women in their 'place' and restricted women's movement so
as to render them incapable of venturing outside their
'sphere.' "[34] Amelia Bloomer may have been better known (and
ridiculed) for her "short dress" but "she was prepared to champion women typesetters against the male printers on her paper [The Lily]," another example of the interrelationship of the
factions.[35]

The arrival of Sarah Pellet, M.D., "a regularly educated
physician and an accepted lecturer on 'woman's rights' and
kindred subjects," in September 1854, marked the formal beginning of "political" talks — and activity — by women in San
Francisco.[36] The account by the *Daily California Chronicle* of her
first address, "Political Reform and the Means of Obtaining
It," not only recorded a noteworthy event but also set the style
for numerous stories about women and their political activities which were to appear in the non-female press in the West
over the next thirty years, the same tone already having been
set in the East.

"Last evening, Miss Sarah Pellet delivered a lecture in the
Musical Hall, on the subject of 'Political Reform, and the
Means of Obtaining It.' By half-past seven o'clock, the hour
announced for the lecture to commence, about one hundred
and fifty persons were present. Of that number nearly one half
were ladies. During some thirty minutes the assembly patient-
ly waited for the lecturer to make her appearance. Then a feeling of uneasiness and weariness came over the people, some
slight noises were made, and Sarah Pellet stepped upon the
platform. This lady seemed about thirty or thirty-five years
old. She had a modest, demure, old-maidenish look, as if she
scorned the fopperies of her sex. Her hair was simply shaded,

and she wore spectacles. She is the first genuine specimen of a female lecturer ever seen in San Francisco, and her opening public exhibition was naturally looked forward to with a little curiosity.

"In the exercise of a sound discretion, we spare our readers a very full report of the lecture... . A little of it was commonplace, and the rest rigamarole. ...She stammered occasionally, read her notes slowly, and altogether was painfully dull. However, the people bore it all with good grace. Many seemed in agony; but they controlled themselves, and took another chew — the men did — while the ladies smiled and yawned. ...

"The whole affair was decidedly dull, and dullness is the worst fault that can be found in a lecturer. Long before Miss Pellet had finished her address, one half of the audience were in a mesmeric trance,...the other half of the assembly moved uneasily in their seats. ... Her last words were...tomorrow evening,... same place, topic: 'True Democracy; Its Definition and Its Illustrations'... ."[37]

The reporter gamely returned for the second lecture and wrote that "At eight o'clock the total number of persons present in the Hall was four. Half an hour later, the audience had increased to twenty-two, by actual count. 'As it is a cruelty to load a falling woman,' we shall say no more on the subject."[38] A rival newspaper agreed on the numbers and went on to say that "This must have been discouraging to the lady's prospects of reform... . We may mention, however, to allay whatever disappointment she may feel at the want of appreciation of her 'mission,' the fact that Mrs. Clarke, who styles herself 'Editor' [as opposed to "Editress"] of the *Contra Costa* newspaper, has devoted two and a half columns to an encomium of her and the 'active promotion of the great philanthropic movements of the day.' So she need not despair of friends and two dozen presumptive converts each night."[39]

The advent of Mrs. Sarah Moore Clarke as editor of the

weekly *Oakland Contra Costa*, also in 1854, marked a beginning for women who became prominent in California journalism and strong advocates of woman's rights in Northern California. According to Edward C. Kemble, Mrs. Clarke was ". . . an estimable lady and accomplished writer. The *Contra Costa* was intended to do service as a ladies' paper, as well as in the drudgery of a general news organ. It was edited with much ability, but only lasted about one year, the editor's ill-health, as well as the limited sphere of support compelling her to retire from the field she occupied so well." The *Contra Costa* was published from 22 September 1854 to September 1855 and was printed in San Francisco at the *Evening Journal* office. Mrs. Clarke also edited that newspaper for a time.[40] She was a notable first among women whose convictions about specific issues may have varied widely, but showed that women had a distinct role to play in California life. Sarah Moore Clarke died, at age 60, on 16 April 1880. Her obituary in the Alta noted that she "... had written with considerable political talent."[41]

A few days after Sarah Pellet's lectures, a follow-up editorial appeared in *The Golden Era* under the heading "Strong-Minded Females." Beginning with the accusation that women had come to San Francisco ". . . from the great hotbeds of transcendentalism at the east — each engrafting upon the tender stock of California her peculiar views of everything connected with our human economy..." to a climactic diatribe that ". . . 'man is a tyrant' — so runs the tenor of their screech-owl song... ." Furthermore, the paper declared, "Not a few of them go so far as to advocate the idea of doing away with marriage as an institution, and having the term of cohabitation dependent upon the will or whim of one or both parties... ."[42] The use of the word "transcendentalism" as a pejorative implied an understanding that Sarah Pellet was a traveling symbol of the kinds of educated women who were the foremost promoters of female causes.

Another of the very earliest woman-edited journals in

California was *The Ladies' Budget*, "published now and then," whose first issue appeared in Yreka (founded 1851, Siskiyou County) on 13 February 1856. Readers were greeted by a declaration that "If women's fortitude is all that is necessary to carry out our purposes, who dares to venture the idea that we shall fail?"[43] The women added that "While we would disdain the thought of being in the 'Woman's Rights' phalanx, we would at the same time rush to the rescue of that refining influence which was given to us by the God of being, and prevent the direful consequences of entire forgetfulness of it among those who are here as strangers upon this strange coast."[44] Kemble called this newspaper "A neat little sheet... under the superintendence of the fair members of one of the churches, and for the benefit of their Society. It only numbered two editions [February, March]...."[45] This small-format paper appeared in a lavish, well-printed typographical dress but there is no evidence that women set type for it.

By 1857, there was record of a woman working in a San Francisco printing-office but the exact nature of her job was not stated. Mrs. Sarah Maiers (A 155) was employed by Whitton & Towne (after September 1858, Towne & Bacon) from the first week in June to her last payday, 18 July 1857. During this time, she appears to have been paid at a rate commensurate with that for typesetters, but there is no proof. However, the rate of $2.50 per day is higher than for trainees or casual labor.[46] Towne & Bacon records show women working passim until the early 1870's but, as with the men employees, their specific tasks were never stated. It is possible to trace a number of men who later were employed elsewhere or running their own printing businesses. Comparisons with their pay scales and work weeks indicate a very strong probability that the Whitton & Towne and Towne & Bacon female employees were indeed typesetters, if only because this was the principal work available to women in printing-offices in this era.[47]

With the ever-increasing number of women migrating

to California after 1848, almost every community gained woman's rights advocates of one stripe or another who frequently had no support structure or print outlet except letters to editors. Matters began to take a different turn, however, with the appearance in San Francisco of a new woman's literary magazine, *The Hesperian*, in May 1858. The editor, Mrs. A. M. Schulz and her assistant, Mrs. Hermione Day, began loftily by declaring ". . . we shall address ourselves to Woman. ... Her home is at once her Eden and her empire, and we should not tempt her to forsake that holy province for the untried fields of fame."[48] This attitude was much like that of the Yreka women, but by the fourth issue (after the departure of Mrs. Schulz) Mrs. Day took aim at a still-sensitive subject: "Look at the employments vouchsafed to our women. How few they are, how scanty, and how worthless. That which no man of spirit will touch is always good enough for a women. Look at their pay, when they labor diligently and faithfully all their lives long, poor creatures, in the hope of saving a little for the day of sorrow — that they may not be obliged to marry, while the bloom is upon their lip, and sunshine is not wholly extinguished from their eyes."[49] "But in this vast arena [of jobs] there is no room for woman, the avenues where she may labor are *few*, and at best *undesirable*."[50]

In fast-developing San Francisco, job opportunities for women expanded rapidly despite *The Hesperian*'s comment. While it was true that employment numerically was in woman's work — needle trades, domestic service, nurse, waitress, etc. — there was no lack of situations. Examination of Directories of the period and the 1870 Census sheets randomly shows the following occupations filled by women: actress, book canvasser, "doctress," book folder, gold beater, "electropathist," artist, paper-box maker, copyist (before the typewriter), photographic retoucher and many others. A number of women were working wives as well. However, the light-manufacturing sector offered limited opportunities because of the competition from Chinese male labor and, in printing, the opposition of male workers and the Union. The question of inferior pay for women also reflected Eastern values. The problem has remained to the present day when one 1990 news story — among many — reported that "newly minted female M.B.A.s earn about twelve percent less than male grads."[51]

With the growing influx of women who needed to work, support groups rapidly appeared. A Ladies' Protection and Relief Society organized in San Francisco in 1853 for two purposes: ". . . to render protection and assistance to strangers, to sick and dependent women and children...[and] women in want of employment can be furnished with appropriate situations."[52] Good jobs for women may have been relatively few, but newcomers, at this early date in San Francisco's development, already had a haven.[53] Not until many years later did specialized training become available, including typesetting. These opportunities came only with the advent of determined woman's rights advocates, who realized that the workplace was the best area to establish the rights of woman, as opposed to depending on suffrage, which the mainstream, Eastern, rights' advocates, and some of their Western followers, saw as the answer to *all* of woman's problems.

The needs of women extended to health as well, and so the Ladies' Union Beneficial Society of San Francisco was organized in September 1860 "to take care of the sick members." According to the by-laws, women from fifteen to fifty could join, pay their dues and be eligible to bring their medical bills for reimbursement. The organization would also contribute $40 towards funeral expenses.[54] The lower age limit of fifteen years was undoubtedly set to enable working girls to acquire coverage. This was not only an early forerunner of medical plans in California, but also another in a wave of all-woman organizations dedicated to single issues, which were to follow in the busy twenty years ahead. Printing would be one of them.

The ratio of the sexes in San Francisco, by 1860, was 25

males for every female. Not until 1880 did this proportion nearly even out, although in the older population, 22–65 years of age, males still outnumbered females.[55] (The 1860 Census enumerated a total of 56,000 residents.) One source explains the disparity in numbers as a concomitant of a mining frontier society while "The scarcity of women laborers posed a serious problem for the development of various light industries in California," a statement at variance with the claim of *The Hesperian*.[56] Although women found employment as typesetters in increasing numbers on the East coast, there is no verifiable record of a woman working at the trade in California until 1857, despite the fact that women potentially represented low-cost workers in an expensive labor market.

Writing in 1882, John Hittell noted that "In the early days, San Francisco was the printer's paradise. Higher wages were paid compositors, and higher prices were charged for work, than was ever known in the world's history elsewhere. In 1870 wages began to decline, owing to the increased number of workmen, who had become so many as to crowd each other, causing competition between firms and in prices."[57] Appendix A reflects the increase of woman typesetters in the work force in San Francisco that began in 1870 and accelerated greatly in the 1880's. By 1882, there were 1,816 compositors at work in California, of whom 200 — or 11% — were females.[58] By 1880, fifteen percent of women aged fifteen and older were working in San Francisco, the majority of them single and young.[59] Eight years later, the city had a total of 771 printers, union and non-union, 195 of whom were women.[60] Of the 195 woman printers, only ten were working in the more lucrative Union-dominated newspaper composing-rooms.[61] The reasons women had difficulty finding work in printing in California until 1870 were twofold: the high wages being paid in San Francisco and the long-standing prejudice against them by both employers and male workers.

In September 1860, *The Hesperian* announced that "Our

49

friends in the city and interior are informed that we are now prepared to do job, book and fancy printing of *every description*, in the most elegant and desirable manner."[62] So "elegant" was the magazine that the Mechanics' Institute awarded Mrs. Day a Certificate of Merit for a copy printed on satin, at the Second Industrial Exposition in 1859: "This paper, ably conducted by Mrs. Day, is published semi-monthly, in quarto size on good paper, and its literary merits are entitled to liberal support."[63] However, there is no record of woman typesetters being employed in the shop at 6 Montgomery Street. While Mrs. Day was traveling in 1862, Mrs. Elizabeth T. Schenck, a prominent leader in woman's suffrage during the next two decades, took over briefly.

An important change for woman printers in San Francisco came in late 1863 with the arrival of the next editor who promptly elaborated on the magazine's history in an advertisement: "Mrs. E. T. Schenck became editor of the 'Hesperian' at a most unfavorable period. It was in debt, in bad repute, small, and struggling for the last breath of life . . . and to her is now due to no small degree the present existing Magazine. In a few months Rev. J. D. Strong and lady became sole proprietors of the work, and in November, 1863, it passed into the hands of its present editor. Mr. Strong made the change of name from the 'Hesperian' to the 'Pacific Monthly,' and gave it a certain popularity, which will ever be remembered, and, we trust, still cling to it.

"The Magazine is now on a permanent basis. It has increased its size to four forms [i.e., four 16-page gatherings], and is circulated to two thousand readers, many of whom are Eastern people. Its aim is to illustrate the life and many interests of this coast, and rank first in worth, as well as age, in the State. LISLE LESTER, EDITOR."[64]

Chapter 4 A Dynamic Personality

Lisle Lester (A 147), one of the ablest and most colorful women who ever came to San Francisco, arrived in November 1863 by sea from New York, to take over as publisher and editor of *Pacific Monthly*. Her extensive background in travel, journalism and printing was most unusual in a woman of this period and her new duties would challenge all of her skills from the outset.

Her given name was Sophia Emeline Walker. She was born in 1837, at Richmond, New Hampshire, where she remained until age six when her family moved to Fond du Lac, Wisconsin, to join Dr. Mason C. Darling, her aunt's husband, and his family. Dr. Darling was an enterprising man, so when he learned that Amos Lawrence, of Boston, had offered $10,000 to start a college if a matching amount could be raised, Dr. Darling took the challenge, and thus Lawrence University, in Appleton, was founded in 1853. It was noteworthy at once because it was one of the few institutions of its time which would accept women.

The following year, Sophia Walker entered, aged 16. Here she found a stimulating environment, ranging from fellow-students to the library with its many books and magazines. The periodicals interested her so much that she decided to become a journalist and as a result never graduated but left Lawrence to join another woman in publishing the first magazine in Wisconsin, *The Badger State Monthly*, "devoted to Art, Science and Literature." Here Sophia Walker published the first of the numerous travel reports she was to write over many years and from coast to coast. While working on *The Badger State Monthly*, she met two of Wisconsin's prominent editors, Beriah Brown and George Hyer. Brown was editor of San Francisco's *Democratic Press* when the new editor of *Pacific Monthly* arrived, and it is probable that he was the middleman who arranged for her to come to San Francisco.[1]

Both her Wisconsin editor-friends had urged her to learn to set type, and this she accomplished as she moved from publication to publication. Meanwhile, she married Isaac L. Bloomer, a teacher in Appleton, 10 October 1858, but sometime in 1861, after an incident concerning her husband and the *Milwaukee Sentinel*, the Bloomers were divorced. In March 1861, Sophia moved to Waupun, in Fond du Lac County, to edit the *Prison City Item* and it was then that she chose "Leisle" Lester as her name for the masthead but became Lisle Lester (used hereafter) when she came to San Francisco.

Meanwhile, the Civil War had started and a shortage of compositors gradually developed in Milwaukee as men went off to war. By 1863, the *Milwaukee Sentinel* responded to women's pleas that they needed work while their wage earners were absent, and Lisle Lester was engaged to teach them typesetting. Although the women were trained away from the men in another building and were paid Union wages, the men objected to women doing this kind of work and struck, thus affording Lisle Lester her first struggle with the problem of women and the Union. She won this battle when the strike was called off, and three months later the newspaper declared that the women's work was superior to that of the men. Lisle Lester had done her work well.[2]

LISLE LESTER
Courtesy New Hampshire Historical Society

James J. Owen, editor of the *San Jose Mercury*, a strong supporter of woman's rights, later described Lisle Lester: "For the benefit of those of our readers who never saw the editress…we will say…that she is a spunky little piece of femininity, weighing about ninety-five pounds avoirdupois, and as chock full of good natural impudence and wit as ninety-five pounds of woman can well be. She is a brunette, not handsome, neither is she particularly plain, and at a rough guess we should say that about twenty-eight springs [twenty-six in fact] have passed over her head! She is unmarried, of course. He must be a bold man who would dare storm the citadel of Lisle Lester's heart, fortified as it is with a columbiad[3] at every embrasure, loaded to the very muzzle with chain lightning. We would sooner lead a forlorn hope over the fortifications of Richmond."[4] In a word, formidable.

"Editress Lisle Lester," as she was first designated on the cover of *Pacific Monthly*, began her work in December 1863 with a rhetorical question: "Will it be a 'Woman's Rights magazine'? Emphatically, No! Yet it is an advocate for the *proper* rights of women… ."[5] Having made this distinction, she pursued her advocacy in the "Editor's Table," a recurring section mostly in six-point type, and in February 1864 plunged deeply into her favorite subjects. The extensive excerpt which follows furnishes an excellent example of both her attitude and hortatory style.

"The subject of female labor, her proper field, and her accustomed remuneration, is brought before the public almost every day in periodicals, newspapers, common conversation until it has become a subject of no inconsiderable discussion, among both sexes. There have been two items, which have occurred under our notice, within the past two months that grants [sic] us the privilege of a few words, in the behalf of females who are dependent upon themselves for a livelihood. It is a matter of earnestness and principle, and we allude to it, not only seriously but with emphasis. In all cities female labor has been limited to a few channels, only such as society and custom, have imposed for years, until those very channels have become overburdened with applicants, and society drawing the lines of limitation so close, that it actually seems to be a matter of great astonishment if a woman dare earn her living at any other respectable business, other than that which has been laid down as her field and destiny. We can hardly pass the subject without giving it a thorough revision, and expressing the feeling and thoughts that arise upon it; but reluctantly we are obliged for want of room, to confine our words to that branch of business which ladies are engaged in to the utter astonishment of a part of society, namely, the *printing business* or more particularly, *Typesetting*. It is astonishing what an undue amount of excitement and discussion is raised over this one fact: namely, a lady *setting type*. People consider her out of her place 'at the case,' but a man is not out of place in a fancy dry goods store! It is singular how people will rule out of place all vocations that [are] proper and well adapted to ladies, and condemn any attempt that one may make to build up the right, as well as the respectability of other channels, for woman's labor. The *art of composition* in a printing-office is perfectly adapted to female help — and experience tells us that female compositors, who make their living at the case as a general thing, excel, both in rapidity and correctness. This we know to be a fact by personal experiences in offices that have a male and female composing department. ...

"Hardly a city in the Eastern or Western states but have ladies in the printing offices, and not only with success but perfect propriety; and now we come to a fact we wish to place before our readers that they may see for themselves the unjust notions and rule of those who persist in keeping a woman out of a printing-office on the ground of inability to give her work, because a Typographical Union controls them and their business. San Francisco is blessed with such a 'Union,' which in its rigidness, its monopoly, and injustice, has full sway over the 'Press' of the city, and brings publishers and Editors to their terms despite justice or principle.

"During the month of September 186[3], a lady who had worked at the case for eleven years in one of the best printing establishments in the city of New York, came to San Francisco in hopes of finding a more lucrative position, where she could support herself with more ease, and less actual hard labor. Like a great many others, she had a low purse, no friends, home, or dependence upon her arrival, and at once applied for a position as Compositor in the Printing Offices of this city. She went to nearly every office, stating her necessity for employment, and what was the result? [W]hy, every office refused to give her employment for the reason 'their men would all leave if they employed a lady;' 'they all belonged to the "Union," and the members of such an honored body could not work in any office with ladies! ...'

"We are a friend to a Printer, and have ever borne that reputation. We would divide the last crust of bread ... with a printer, but we *will not favor* or tolerate a 'Union' that excludes the right of an honest living to another, just because she is a woman, or to a man, if he finds himself obliged to work for less than Union prices to keep from starving. ... We know of one office that employs a lady Compositor, and a gentleman is employed as foreman — but he does not belong to the 'Union' — and that office is the office of the *Pacific Monthly*; and if we are obliged to change our foreman and engage a man that objects to work in an office that employs ladies, he will get his discharge at once; and if we cannot find a printer man enough to have a little independence about him, and is so afraid of the 'Union' that he dare not work in the office with a lady, we will send East for a Printer, and do the work ourselves until he arrives — as we understand the business well enough to do it, *if necessary*. It is very consoling to feel we are competent to control our own business independent of any tyrannical monopoly."[6]

The Typographical Union, which Lisle Lester battled so

forcefully, went through four stages before it became a permanent fixture on the San Francisco scene. The Pacific Typographical Society, formed in 1850, lasted only three years, principally because of the fluctuating nature of the labor force during that time. Its second version, the Eureka Typographical Union, continued from 1853 to 1859 when a combination of social and economic conditions — including a depression — in San Francisco caused it to fail. The reconstituted organization, Local No. 21 in the National Typographical Union, retained the title, Eureka Typographical Union. It is this version of the Union with which Lisle Lester had her problems because by 1864 the reorganized group exercised more control over the industry than its predecessors. The typesetting scale, from 1859 until a big strike in 1870, was stabilized at 75¢ per thousand ems and Lisle Lester adhered to the Union figure, as she had in Milwaukee. After a strike in 1870, the Union failed one more time before it stabilized into a permanent organization.[7]

Lisle Lester's anti-Union salvo had an interesting response in the *Daily Morning Call*: "The February number of [*Pacific Monthly*] magazine has been received. Its talented editress, Lisle Lester, displays good judgment and fine taste in her work, and the magazine would be quite readable, were it not for the typographical blunders and orthographical errors so thickly strewn on every page. In one article claiming superiority for female over male compositors, three or four errors occur in as many lines, and, if — as we presume — the matter was typed [set in type] by a lady, tend very materially to defeat the force of the writer's argument. It is a pity the editress' skill should thus be rendered ineffective."[8]

The *Call's* assessment drew a testy reply from "A Female Compositor" which the paper refused to print because "they wished to see the lady first." *The Golden Era* did publish it[9] and the letter was then reprinted in full in the March issue of *Pacific Monthly* with a lengthy commentary. The female compositor said she had ". . . arrived in this city a few months ago, a stranger, friendless, homeless, and without means. I called on you for employment as a compositor, and met with a cold reception . . . . But I was not 'sore,' and went away and should not have replied except for this last unjust attack on me in your journal." She continued by reciting the woes of women attempting to get any kind of employment in San Francisco and cited the fact there were few "manufactories" and those "principally employ Chinamen. . . . 'Lisle Lester' has taken a stand and dare[s] maintain it!"[10]

Ms. Lester rejoined that she could and would conduct her business as she pleased and refused to be under any monopoly — the Union again — and asserted that the public would verify her words "to the extent and influences of such a tyranny." She named the female compositor as "Miss McKee" — *twice* — but the only record of a female compositor at work at this time shows a Miss Mary McGee (A 163), who also appears as the president of the Female Typographical Union announced by Lisle Lester in May 1864.[11] This confusion in names may have resulted from typesetting (or proofreading) of the kind derided by the *Call*.

Ms. Lester continued by declaring that "...our office rejoices in the addition of another lady Compositor, and when our Printer arrives from the East, and we turn on the steam on our own Press, under our own shingles and siding, we shall extend an invitation to our friends to call and see if the plan is not as feasible and good as any other." The second female was Mary E. Parker (A 201), an 18-year-old apprentice typesetter. Ms. McGee, Ms. Parker and Lisle Lester herself are the first women in San Francisco who are specifically identifiable as typesetters and the sparse forerunners of the compara-

Fig. 9, facing page: *Revised cover design for all issues of The Pacific Monthly (copy at bottom varied) under Lisle Lester's editorship. Said Lisle Lester, "Certain it is very chaste, and so new, in its design, and odd in its construction, that it really deserves notice."* Courtesy Robert J. Chandler.

MARCH, 1864.

VOL. XI.

No. 3.

THE PACIFIC MONTHLY

EDITED BY

LISLE LESTER.

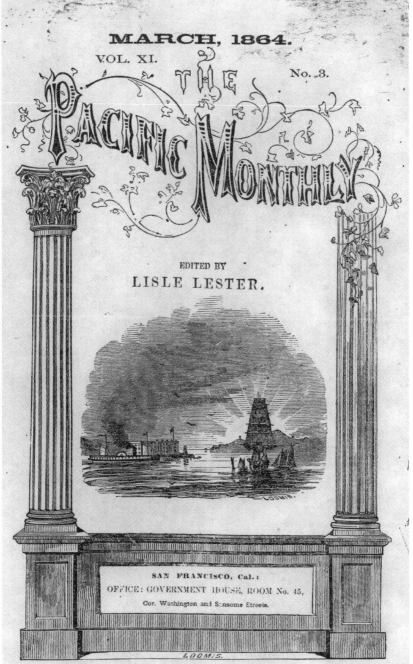

SAN FRANCISCO, Cal.:

OFFICE: GOVERNMENT HOUSE, ROOM No. 45,

Cor. Washington and Sansome Streets.

We Circulate ONE THOUSAND COPIES *Monthly.*

TERMS, $3 00 per Annum ; $2 00 for Six Months.

tively large number at work throughout Northern California by the 1880's.[13]

The aroused editor of the *Pacific Monthly* had not finished and said she understood that there were enough female type-setters in the state "to form a quorum" for establishing a union of their own. She could not resist one last swipe at the men's Union. "We hope the 'Typographical Union' has no law or power to prevent the establishment of a 'Female Typographical Union,' and that they will prosper in its establishment, for we love to see the fun, and opposition is the life of business."[14] Clearly she did not think this idea feasible for San Francisco alone because of a scarcity of female typesetters and hence the mention of a "quorum." Most importantly, no such strategy had yet been tried anywhere else. The widely known Woman's Typographical Union #1, created parallel to the all-male Typographical Union Local #6 in New York City, was not established until 1869.

Meanwhile, Robert Montgomery, editor of the *Napa Register*, after noting that the *Pacific Monthly* had been using female compositors, commented: "Well, this is an innovation upon ancient usage — but what of that? Ladies need employment suited to their strength, and why is not type-setting as suitable as sewing? — If there is any starving to be done, men can stand it better than women, and can shift their location and employment with greater facility. We have two daughters and intend, if they live to a proper age, to make them both accomplished compositors. A liberal course of types will teach them many things better worth knowing than how to murder French or torture a piano. If they have any brains they will be proud of being both independent and useful; if unfortunately they have not, we can restore them to the customary employment of missdom — working red worsted dogs, shattering nonsense and spoiling water-colors."[15]

Lisle Lester meant what she said and announced in her May issue that "A Female Typographical Union was organized

April 25th 1864 in this city [San Francisco]. Miss Mary M'Gee, President; D. S. Cutter, Secretary; Miss Emma Goodale, Treasurer; Lisle Lester, Mary E. Parker, Mrs. E. T. Schenck, Wm. B. Ludlow, Executive Committee; Mr. Tenat, Geo. Sprague [C 9], and Mrs. F. G. M'Dougal, Committee on Rule and Finances.

"This 'Union' is established for the protection of female compositors and gentlemen desiring to attain employment in an honorable manner. The By-Laws of the 'Union' will be printed in a few days. The 'Union' was organized *with a large membership.*"[16] An out-of-town newspaper furnished a summary of this action, reporting that there were "...over thirty ladies and fifteen gentlemen as members.[17] All the gentlemen are practical printers, and several of the ladies have worked at the 'case' in the Eastern States for some years, the remainder of the ladies intend learning the business." The writer also could not resist adding that "We don't wish to be ungallant, but think it a bad business for printers to be engaged in encouraging such innovations in the profession."[18] No copy of the By-Laws has been found to date. The organizers named included one male compositor, George Sprague, who is listed in the 1864 Directory as working at *Pacific Monthly*, and the three women already prominent in the magazine's affairs. No others have been identified as having connections with printing. The "large membership" appears to have been a contrived affair, but Lisle Lester and Mary McGee had indeed been typesetters in the East. The "quorum" appears to have evaporated altogether, and the question of female typesetters elsewhere in the state was left moot.[19] The relation of others to printing appears to have been minimal. For example, Mrs. Schenck's previous editorship of *Pacific Monthly*, frail health and a continuing career on behalf of woman's rights, made her an unlikely prospect for apprenticeship in typesetting, while Mrs. Fanny G. McDougall, who wrote more than a half-dozen articles for *Pacific Monthly*, was preparing to put out a Spiritualist publication in Sacramento.

More pressing matters supplanted Lisle Lester's promotion of her Female Typographical Union. In her June issue, she offered "An apology to our friends, for our tardy publication, and the typographical errors [finally admitted publicly!], which are found in part of the book.[20] Our delayed visit to Nevada, and a general recharging in our office, has caused our delay and brought upon us the duty of apologizing for the errors. We have secured the services of two excellent female compositors and hereafter the 'Pacific Monthly' will appear promptly and in readable shape."[21] Apparently, she terminated the old crew — or they left — and thus the core of the Female Typographical Union was gone. A hint appeared two months later in the obituary of Mary E. Parker, who was killed 5 August 1864 in a flour-mill accident in Solano County, which said that she "was for sometime [sic] in the office...learning the art of type-setting... ."[22] The past tense suggests that Ms. Parker, one of the Female Typographical Union founders, was not employed at *Pacific Monthly* at the time of her death.

The typesetting and proofreading were improved in the July issue of *Pacific Monthly* although gaucheries, such as "two" for "too," and some vagaries in spacing and punctuation persisted. But the redoubtable Lisle Lester was at the top of her form in assailing the men's Union once again. She thought that "The pages of a literary magazine are too good to be converted into vehicles of controversy and combat... ." Nevertheless, she filled three pages of small type with confrontation and stridently attacked the Union's claim that "women work for less wages. ..." "Any man who will attempt to prejudice the public against female help in printing offices by bringing forward such a false assertion, and such a weak, brainless, puny argument as that, must be either a fool, or the double-refined quintessence of mortal meanness. In such a small, caviling, detestable excuse, is a clear evidence of a painful want of brains, and a common respectability of principle. No such argument could originate from anything honorable, upright, or manly.

"The reason is both absurd, and destitute of the smallest grain of honor. The second reason, 'that ladies cannot do the work as well' is still more ridiculous, and another witness to the blockhead intellect, that patched it up. ... The whole argument is puerile, low and disgusting, far beneath the dignity of a true gentleman to cherish, and by far two [sic] contemptible to give to public utterance, and unworthy of any association. It gives this 'Union' no credit, to attempt to strengthen its platform with such nonsense and petty objections, it clearly defines, beyond all misrepresentation, or mistake, a usurping power, a ruling monopoly, a selfish organization which is gently and politely forced into the opinion and tolerance of a public, under the skimmilk [sic] insinuation of a 'Protective Union.'"[23]

Finally, to support her arguments, Ms. Lester described at length her problem in attempting to replace her foreman who had become interested "in a new coal mine" and had left.[24] Upon her return from another trip to Nevada, she was "much pressed with hurry and labor" and asked her "publisher" to send over a man to "'make up forms' for press."[25] The response was that all the men belonged to the Union and "they do not wish to tolerate female compositors." Ms. Lester still had another page to go on this subject and concluded by reporting that "We have made up the said forms ourselves and feel perfectly competent now to transact our little business without calling upon the kindness of 'Union' men who are *afraid* to be found in an office that employs women."[26] None of the woman activists who followed Lisle Lester to San Francisco ever matched her colorful prose style or her vigor for combating the opposition.

The contentious editor always had time and energy for literary recitals and an interesting variant in her audience appeared when the *Alta* reported that "Lisle Lester will give a public reading at the Methodist Church (colored) on Stockton Street on Monday...for the benefit of the church. The program will consist of choice selections from popular authors... ."[27] More accurately, this was the African Methodist Episcopal

Zion Church (organized 1 August 1852) which was on the West side of Stockton St., between Clay and Sacramento Streets. It had a library of 400 volumes which indicates a literary interest on the part of members.[28]

Lisle Lester's reputation evidently had few bounds and her non-traditional views made her interesting not only to groups but also individuals. When "An Old Subscriber" wrote from Red Bluff for a photograph because "we world [sic!] like to see the Editress," she replied that the letter "caused such a heavy jump of our egotistic self-assurance, against the tender panels of our much-flattered heart, that we read, re-read, read again and inwardly felt our goodness... ." While declaring that she "had an aversion to keeping pictures 'on hand for distribution'" and until she could find an artist who will "warrant" a pretty picture of her face which she could send far and wide, she would decline the request. She referred the writer to a likeness of the Goddess of Liberty which could be easily found on all U.S. coins and to which she always imagined she had a "striking resemblance."[29]

Lisle Lester may engender sympathy for her battles with the Typographical Union but the ultimate failure of her magazine cannot be attributed solely to Union activity or male chauvinism. Her personality also contributed to its demise. She was an entrepreneur in a man's world and, because she was not a self-contained enterprise, she was always dependent upon the all-male firms for presswork, binding and probably much composition. Charles F. Robbins & Co., 417 Clay St, printed her first issue in December 1863. Turnbull and Smith, Book & Job Printers, at the *Daily Morning Call*, 612 Commercial St., produced the January 1864 number.[30] Neither firm had women employees of record.

There is no information available about the printer(s) for the next two issues, but the records of Towne & Bacon, one of the largest San Francisco printing offices of the time, show it billed *Pacific Monthly* $33.00 for "press work" in April.[31] On 25

THE PACIFIC MONTHLY
— IS A —
LITERARY MAGAZINE,
DEVOTED TO THE LIFE AND LITERATURE OF THE PACIFIC COAST,

AND EDITED BY

LISLE LESTER,

A Lady long connected with the Press of Wisconsin   This is the first and only Literary Magazine on the coast.   Its contributors are among the *First Writers* of America, and its many attractions are such as to insure its future success as the first Magazine of the State.

Its continued series of articles are of a Botanical, Geological, and Descriptive character, confined to California and this Coast.

TERMS.

Per year, in advance $3 00
Per year, if payment be delayed three months... 4 00
Single copies.............................. 25

OFFICE — Corner of Sansome and Jackson, in Thun's Building. Rooms, 18 and 19.

Fig. 10: *Lisle Lester advertises her magazine.* Courtesy Robert Chandler.

April there was another entry, "Lisle Lester, 500 due bills, 'Pac Monthly,' $3.00, 100 posters, $4.50."[32] The relationship continued into the next month when on 11 May "Pac Monthly — Lisle Lester, No. for May, $18.00." is entered.[33] There is no record for June, but Towne & Bacon posted another $18.00 for producing the *Pacific Monthly* in July.[34] No further account for printing occurs, but in August the following cryptic entry appeared: "By amt. paid July 21 $12.50" and in the next column "$83.25," which was probably Lisle Lester's outstanding balance.

Trade presswork at this time was not expensive, as shown by a billing to Towne & Bacon, dated 30 December 1865, which lists "use of Press 10 hours" at $2.50 and other specified presses, all of which were charged at 25¢ an hour.[35] Being in arrears, as suggested by the Towne & Bacon journal entry, may have indicated financial problems. However, Lisle Lester's magazine consistently had advertising, varying from a low of 4 $^1/_2$ pages in January, her first month in full control, to a high of 10 pages in October, so there should have been income in addition to the subscriptions. Unfortunately, business records for *Pacific Monthly* do not exist, but a reasonable assumption is that her paid circulation was far less than the 2,000 (@ $3.00 per year) claimed on the cover each month. Her order for the 500 "due bills" from Towne & Bacon in April may have been indicative of reality.[36]

Lisle Lester's frequent travels out of town left her journal without leadership. When a crisis arose the magazine was late or, in the case of the September number, never issued. One serious consequence of the September omission was that the October issue repeated most of the August pagination, a sure sign shop matters were out of control. To make matters worse, in October page 807 became 808 on a right-hand page — *recto* — until page 817, a left-hand page —*verso* — which is followed by 819, a recto, and pages are numerically and physically correct thereafter.

When she was in the office, she often did not have the time or inclination to be precise. The October issue had an interesting explanation of her production difficulties and is quoted unedited: "To stop the continued questions, of 'what makes the *Pacific Monthly* so late' we have to give the reasons in plain English. ... The first reason is that the printer who had the press-work to do, was determined to have his pay in advance, for the work, an accommodating piece of kindness he had never dared to lavish upon any one else — and because we wouldn't give it, he kept the material for awhile — and when it came home, he felt gentlemanly in making it as near hash as possible, to make us more work — then, the other press man, all at once discovered that his press had taken a fit, and would only do 'newspaper work;' something quite funny for a press that had been doing the heaviest book-work in the city, to so magically change; quite a sweet scented bouquet of just such performances have detained the 'Pacific Monthly,' against which the 'Union,' use their best endeavors to crush — it has stood the storm so far, and in all probability will continue to do. Unless the 'Union,' take the throne as Executive Committee of the whole to govern the world at large."[37]

Lisle Lester's shot-gun approach to her problems raises questions. First, how was the Union actually involved with any specific problem? Second, how did she know that the firm from which she sought presswork had "never" before required payment in advance from anyone else? Third, how could she have expected a rush of co-operation after her vituperation of men printers in previous issues? Fourth, was it not possible the press needed to print her magazine at the second printing-office might have had other work scheduled at the time she was demanding her work to be done? Lastly, how could the printer be blamed for the inconsistent style in the typesetting in the quotation above — to say nothing of the wonderful run-on sentence that is its core?

On the surface, Lisle Lester's problems do not appear to be Union-related. A more likely explanation is that she either still owed a large balance on work previously done, as in the case of Towne & Bacon, or had been slow paying her bills, or both. The time taken in returning her typeset materials may have been a matter of opinion concerning what constituted "awhile." However, there is some validity concerning the "hash" that she had to correct. The October issue has one surviving example in the full-page advertisement for "The New England Family Sewing Machine." It had been running regularly, and now came back to her in disarray and was never corrected. A "y" at the end of one line, a "c" at the start of another and the following eight characters, are missing altogether while the last line has been moved to the right two picas.

The eleven issues of *Pacific Monthly* under Lisle Lester's

management show strange inconsistencies in typesetting. Many of the articles were set as well as any in San Francisco at the time and showed few typographical errors (only battered types here and there). Her "Editor's Table" section frequently exhibited poor proofreading on the one hand, and some unskilled spacing and strange use of punctuation, especially commas, on the other. Putting in commas where they were clearly not needed may have been an amateurish way to justify a line quickly and not show too much white space between words. Much of the poor work was done by her changing crews of female typesetters. Given her temperament, Lisle Lester may also have been consistently late with her "Editor's Table" copy, while the pressure of her deadlines may have led her to minimize the proofing. With all the typographical errors, what was the condition of the proofs that were in the strings proffered for payment? The basic rule was that only corrected proofs were pasted up and submitted.

Lisle Lester was again out of town when her staff prepared the November issue, and it was "unavoidably postponed." When it finally appeared, two sentences literally read as follows: "Tht notice says, that the discourse and services, were highly interesting ; which must be a compliment to Mr. Haswell. The Flag . comhliments the absence of the discourse, making the matter rather ludicrous; wit is the soul of ———."[38] Not surprisingly, this was the last issue.

The coup-de-grâce to Lisle Lester's endeavors came as she was traveling on a reading tour in Oregon that November. The stagecoach in which she was riding had a serious accident, and the injuries she sustained kept her from working further on the magazine. When she had recovered, she tried to raise money by giving readings all over California during the spring of 1865, but she never was able to gather the funds needed to revive the *Pacific Monthly*. However, she clearly had devotees everywhere for while she was visiting Grass Valley (Nevada County), the local band came one evening to her hotel to play a serenade "as one of the pleasant surprises the town of Grass Valley is capable of giving to its friends."[39]

After finding it impossible to revive her magazine, Lisle Lester again began to travel and the next notice of her in California was an 1867 news article which reported that "Mrs. L. P. Higbee [A 123], alias 'Lisle Lester,' recently gave a reading at Silver City, Idaho. It would appear she has captured a husband somewhere in that region."[40] Indeed, the marriage of Lyman P. Higbee and Lisle Lester had taken place on 29 August 1866 in Fond du Lac, Wisconsin. Mrs. Higbee had first known her new husband in Richland Center, Wisconsin, when she was editing a magazine there and he was practicing law nearby. Following their marriage, they went to Idaho and lived in Silver City.[41]

She reappeared in California in 1867 when the *Alpine Miner* (Monitor, Alpine County) reported, "The entertainment, given by Lisle Lester . . . was extremely well received. Her 'readings' are pleasing and entertaining with nothing of the prosy about them. Mrs. Higbee (the lady's real name) is collecting statistics for a book, which she is preparing for publication, and which we antisipate [sic] will be a work of no ordinary merit."[42] There is no traceable record of the book.

Lisle Lester Higbee had returned to San Francisco by 1868 and appeared in the 1869 and 1870 Directories as "Mrs. Lisle Higbee, furnished rooms, 313 Jessie St." She had no listing in the 1871 Directory but there was an entry for "Liman [sic] P. Higbee. attorney and counselor at law, 313 Jessie St." He was also entered under "Attorneys" but never appeared in a Directory again because sometime thereafter he returned to Idaho without his wife.

The San Francisco to which Lisle Lester returned was changing dramatically from the time when she was editor of *Pacific Monthly*. Emily Pitts Stevens had taken possession of the *Pioneer* and was becoming a force in promoting woman's organizations and affairs, including the employment of women in printing. Ms. Lester's second stay also coincided with the sensational trial of Laura D. Fair for the murder of her married lover in November 1870. Her trial in 1871 became a rallying point for the "strong-minded females" because they regarded her treatment at every turn as prejudicial, starting with the selection of an all-male jury from the voter rolls from which women were excluded. Thus, they said, Mrs. Fair was not tried by a jury of her peers but she was nevertheless sentenced to be hung, the first woman to receive the death penalty in California. The day after a stay of execution was issued, Susan B. Anthony and Elizabeth Cady Stanton arrived in San Francisco on a speaking tour. These femininist leaders soon joined the local women in visiting Mrs. Fair in her cell regularly, and they also spoke out forcefully on her behalf, thus creating much hostility in the community.

Lisle Lester, as Mrs. Higbee, joined these visits and was threatened for her support of Mrs. Fair. Matters took a serious turn one night after she had been visiting Laura Fair's mother. She was attacked as she descended a staircase by a large man wielding an umbrella, the handle of which was found beneath Mrs. Higbee's head where she had fallen senseless. The *San Jose Weekly Mercury* commented that "It is eminently characteristic of the San Francisco press, in referring to this affair, that they should affect to disbelieve that any assault was committed, and that the thing was a put up trick to create public sympathy for Mrs. Fair. The *Chronicle* says of it: 'Better sell such stories "short" — in broker parlance.' *The Call* speaks of it as an 'alleged mysterious assault.' *The Bulletin*, more honest, is nevertheless inclined to be incredulous, although apparently willing to deal justly in the matter when convinced. And all of this is in the face of the certificates from two physicians, to the effect that Mrs. Higbee was suffering from 'concussion of the brain resulting from a blow upon the head.'

"We sincerely hope the ruffian may be discovered, and full and complete justice done in his case."[43]

Mrs. Lane, Laura Fair's mother, offered a $5,000 reward

for the apprehension and conviction of the assailant. In August, the *Chronicle* further reported that Mrs. Higbee had recovered enough to walk all the way to Mrs. Fair's cell to continue her visits.[44]

Mrs. L. P. Higbee was listed as a "printer" in the 1872 Directory. At a time when there were many more opportunities for women in the printing business than when she arrived in 1863, she had no known connection with any printing-office. Writing from Baltimore in 1877, she said, "I came east three years ago suffering from inflammatory rheumatism to such an extent my hands could not use my means of livelihood, pen and ink," which indicates she was probably unable to set type either. Apparently the climate in San Francisco was troublesome to her condition because, she continued, "I then went to the West Indies in search of some healing spot" and regained her health. Especially intriguing is the heading of the Baltimore letter: "Ladies' Printing Union…Lisle Lester, Manager."[45] Women in printing had become her cause once again.

With her health restored, Lisle Lester successfully resumed her journalistic career in Eastern cities, including Chicago, Baltimore and New York where she died of pneumonia 25 June 1888, aged 51.[46] She had been a dynamic, talented editor and writer and deserves to be remembered for her pioneering — if sometimes abrasive — work on behalf of women in the printing business in San Francisco.

Fig. 11: *Early advertisement showing original address of the Women's Co-operative Printing Union. From* The Traveler's Guide, July 1869. Courtesy The Bancroft Library.

Chapter 5 Women Gain a Foothold

When Lisle Lester closed her *Pacific Monthly*, woman typesetters in San Francisco were as scarce as before her arrival. The 1865 Directory shows her former employee, Mary McGee, working at the *Christian Advocate*, a position she must have taken after the shakeup at *Pacific Monthly*, and before October 1864 when the Directory went to press. There could have been other women in printing-offices in San Francisco at this time but they left no traces, not even in the Towne & Bacon Records. It is only in the outer regions of the state that females in numbers were setting type from 1865 to 1868.

In those areas necessity sometimes dictated the training and employment of women, an example being *The Alpine Miner*, in 1867. The compositor quit so the editor had to "depend solely upon our three little girls to 'set up' the matter for the *Miner*... ."[1] This move is reminiscent of the household period of printing. The following year, another newspaper said that "there are in [Alpine] county about twenty girls who are compositors, and of this number five are employed on the county papers."[2] Further evidence of women workers in peripheral ar-

Fig. 12: *One of the incorporators of the Women's Co-operative Printing Union was J. R S. Latham, an official at Wells, Fargo's Montgomery St. office.* Author's Collection.

74

eas came in an 1868 editorial which said, "One of our country exchanges speaks of a lady who sets type as a 'typoess.' Why not say printress and have done with it? . . . and if she should arrive at the dignity of the direction of an office, would she be forewoman?"[3] The questions were not moot because San Francisco was about to acquire a long-lived printing-office run by and for women.

The turning point came with the arrival of Mrs. Agnes B. Peterson (A 205) and other woman typesetters from New York, by way of Panama on the P.M.S.S. *Constitution*, at 11:00 a.m. on 9 July 1868.[4] She later wrote, "I arrived here . . . expecting to maintain myself and child by obtaining work as a compositor till I could establish myself as correspondent. I found the proprietors of all the offices quite willing to give me employment, but the Typographical Union refused to permit me to work, the members threatening to leave any office where a lady might be employed. I called on the President of the Union, stating in what embarrassment it placed me, having myself and child to support, without friends or money and a stranger. I did not wish to work for any less but asked for an exception to be made in my case, or to admit me into the Union. My request was treated with contempt."[5] Her letter virtually duplicated Mary McGee's earlier protest to the *Call*; Mrs. Peterson's experience confirmed that the same situation existed that Lisle Lester found on her arrival five years earlier. Of special interest is Agnes Peterson's comment that owners wanted to give her employment. The *Bulletin* also pointed out that proprietors were not only willing to hire women at Union wages, but that "the Union is a Society of printers who govern and dictate who shall and who shall not be employed . . . and its members decided female labor shall be rejected." Jobs were not available as a result "even in instances where they needed more help than they could obtain from the ranks of good male compositors."[6]

Because newspapers customarily only listed first-class passengers, assumptions must be made regarding other wo en who arrived with Mrs. Peterson. Two of them were undoubtedly Lizzie V. Harty (A 117) and Emma Hutchinson (A 128) because they appear as working with Mrs. Peterson in the 1868 Directory.[7] The *Bulletin* referred to them as "a number of other respectable women" when reporting their rebuffs and in conclusion said, "Mrs. Peterson has obtained the necessary capital to fit out a complete job printing office,

Fig. 13: *An early example by women showing the style and kind of work common in San Francisco at the time.* Courtesy The Bancroft Library.

to be managed by herself and associates as a Woman's Coöperative Printing Office. Presses, type, etc., have already been secured, and an office has been rented in the third story of the building No. 517 Clay street."[8] This was a major accomplishment and the firm opened for business on 10 August.[9] It may have been significant that the new enterprise was in the same building that housed the friendly *Bulletin*, as well as a leading San Francisco printer, Edward Bosqui. The company was organized quickly enough to take a half-page advertisement in the 1868 Directory that went to press in October and designated it as the "Women's Co-operative Printing Union, Agnes B. Peterson, Proprietress."[10] (Hereafter, WCPU.) This is a contradiction because a co-operative never has a single owner.

The formation of the WCPU in August 1868 represented the first permanent foothold for woman printers in San Francisco and had several repercussions. In the most important, the *Los Angeles Star* not only devoted a thoughtful editorial to the WCPU but also took action: "We have an application from a young girl to be allowed to learn the printing business in our office. We did not encourage the application of the child, yet did not absolutely discourage her. Since the report of the Unionists [WCPU] in San Francisco reached us, we have changed our mind on the subject, and now give her all the encouragement in our power. She devotes her forenoons to the study of modern languages at school and her afternoons, for a couple or three hours, are spent at case, learning the 'art preservative.' Should any of her young companions desire to follow her laudable example, to learn a business as easy of acquisition as sewing, and much more remunerative, we shall be pleased to aid them, believing that type-setting is as much in the line of a woman's vocation as any branch to which she has hitherto been confined by the tyranny of society, and a thousand times more becoming than seeing men engaged in selling ribbons and laces and hoop skirts, or in teaching the young their a, b, c. A printing-office is just the place for woman."[11] The anonymous "girl" may have been the first woman typesetter in Los Angeles. More important was the attitude of the *Star* in its willingness to hire and train others, which was truly noteworthy. There was no Union in Los Angeles to interfere at this time.

In a letter to *The Revolution*, Agnes Peterson elaborated on the start of her new organization: "On making the particulars known, parties advanced the capital and a nice printing office is

77

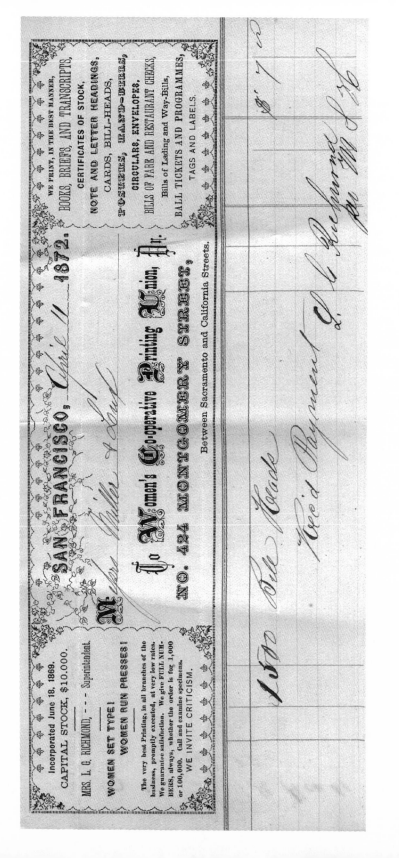

Incorporated June 18, 1869.

CAPITAL STOCK, $10.000.

MRS. L. G. RICHMOND, - - - Superintendent.

WOMEN SET TYPE!
WOMEN RUN PRESSES!

The very best Printing, in all branches of the
business, promptly executed, at very low rates.
We guarantee satisfaction. We give FULL NUM-
BERS, always, whether the order is for 1,000
or 100,000. Call and examine specimens.
WE INVITE CRITICISM.

SAN FRANCISCO, April 11 1872.

Mrs. Miller & Co.

To Women's Co-operative Printing Union, Dr.

NO. 424 MONTGOMERY STREET,

Between Sacramento and California Streets.

WE PRINT, IN THE BEST MANNER,
BOOKS, BRIEFS, AND TRANSCRIPTS,
CERTIFICATES OF STOCK,
NOTE AND LETTER HEADINGS,
CARDS, BILL-HEADS,
POSTERS, HAND=BILLS,
CIRCULARS, ENVELOPES,
BILLS OF FARE AND RESTAURANT CHECKS,
Bills of Lading and Way-Bills,
BALL TICKETS AND PROGRAMMES,
TAGS AND LABELS.

1500   Bill Heads $7.50

Rec'd Payment   L. G. Richmond

pr M J F

started under the head of Women's Co-operative Printing Union. We have the best New York material and presses, and are promised plenty to do . . . ." In conclusion, she said, "However, we do not know anything about or wish to take any stand in politics, or want to vote, therefore you may have a contempt for our lack of strongmindedness. We all strongly sympathize with women who are unjustly oppressed and shut out from an opportunity of earning their living and will be willing to aid any way we can any who have been thrown in the same position as ourselves." A postscript noted, "All departments are to be filled by ladies."[12] Nevertheless, Mrs. Peterson's attitude was another unfortunate example of the failure of working women to understand the connection between their status and political action.

Among others, the *San Jose Patriot* commented forcefully on the rebuff of the San Francisco women: "The Printers' Union took alarm and denounced the employment of females in printing offices, although these young girls did not propose to labor for less than prices fixed by the Union. Although some of the printing offices were in need of help, the *Bulletin* says that they did not dare to give these girls employment against the edict of the Union. What a shameful tyranny this is and how unmanly and disgraceful it was to the printing establishments to pay any regard to the orders of their arrogant masters.

"We are glad to learn that the lady printers, not discouraged by the unkind and discourteous conduct of the craft in San Francisco, hired a room, bought presses, type and other printing material, and are about to begin job printing on a large scale. We mistake the character of the business men of San Francisco, if the ladies['] printing office is not speedily crowded with orders for work — and we hope the Printers' Union and their servile slaves — the big printing offices — will soon have cause to regret their illiberal and ungenerous

Fig. 14, facing page: *This billhead also appeared on purple paper and in various colors.* Author's Collection.

WOMEN'S

CO-OPERATIVE

Printing Union,

LAW PRINTING.

INSURANCE PRINTING.

424 MONTGOMERY ST. 424

Low Prices and Good Work

Are the leading features of this Establishment.

EVERY DESCRIPTION OF

BOOK & JOB WORK

NEATLY EXECUTED.

Orders from the Country specially attended to.

ESTIMATES FURNISHED.

MRS. L. G. RICHMOND, ... Superintendent.

conduct."[13] (Lisle Lester could not have said it better.) Another far-away Southland newspaper also gave firm support: "Open, then, the composing room, invite her to enter, instead of harshly excluding her. Give her work; she can do it all better than half the clumsy boys and men who are now engaged in the business."[14] The WCPU had strong support from the outset and exerted an influence Mrs. Peterson and her colleagues could never have imagined.

Although formal woman's rights organizations had not yet arrived in San Francisco when the WCPU was established, other special-interest groups had been making progress in organizing support for the ever-growing number of women who were migrating West. The Female College of the Pacific was incorporated 17 January 1867 followed, on 27 March 1868, by the Women's Co-operative Union "to furnish employment for its members and pay them as much as it is possible to pay them for their labor, by deducting merely the absolute expense of purchase and sale of the article manufactured from the proceeds of its sale . . . ."[15] According to one account,". . . it is not confined to underwork [sewing]. It embraces photography, painting, shell-work, all in art that a woman can do, even publishing a newspaper or magazine, or writing and publishing books."[16] As matters developed, most of the women worked at jobs that were not quite so glamorous or ambitious, mostly sewing. Occasionally, confusion between this group and the WCPU occurs in catalogues and texts. They were unrelated organizations.

On the medical side, the San Francisco Lying-in Hospital and Foundling Asylum was incorporated in April 1868 for the "care, protection and proper treatment of respectable married and unprotected single women, with their offspring." The California State Women's Hospital, founded in 1867, and incorporated in May 1868, was for "the treatment of diseases pe-

Fig. 15, facing page: *The wood engraving is by a woman, Lila Curtis. Note the press is treadle-operated.* Courtesy The Bancroft Library.

San Francisco, *May 1st* 1876

Mo. Savage Mining Co.

Bought of HASTE & KIRK,

IMPORTERS OF AND DEALERS IN

CUMBERLAND, LEHIGH, AND ALL OTHER BITUMINOUS
And Anthracite Coals; also, Hard and Soft Iron,

Beale Street, bet. Mission and Market.

PAYABLE IN U. S. GOLD COIN.

Women's Print, 424 Montg'y St. S. F.

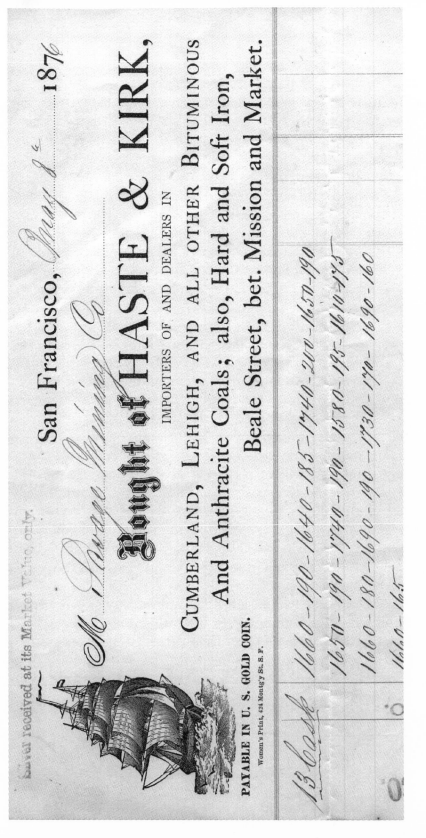

13 Cash	1660 – 190 – 1640 – 185 – 1740 – 200 – 1650 – 190
	1650 – 190 – 1740 – 190 – 1580 – 195 – 1610 – 175
	1660 – 180 – 1690 – 190 – 1730 – 170 – 1690 – 160
	1610 – 165

culiar to women." Clearly, the growing population required services that did not exist. One that was especially useful to women was the California Labor Exchange, incorporated 27 April 1868, whose objective was "to obtain employment for all classes of men and women." From the date of incorporation to 1 July 1870, "employment had been supplied to 24,581 persons, 6,726 of whom were females," mostly unskilled and semi-skilled labor.[17] The arrival of Agnes Peterson and her associates coincided with a time when women were becoming a significant part of life in San Francisco and California.

Concurrent with the organization of the new support groups for women, a Spiritualist weekly newspaper, *Banner of Progress*, published in 1867 and 1868, contained news about woman's rights in every issue. In December 1867, there was a long report on "Womanhood Suffrage in the National Spiritualist Convention at Cleveland" and it covered at length a suffrage speech by Parker Pillsbury, who wrote regularly for Susan B. Anthony's newspaper, *The Revolution*.[18] Of special interest is a report by Lisle Lester on the advancement of "the cause" by Spiritualist lecturers in Nevada and she complimented the *Banner of Progress* as "a shining beacon on the Pacific Coast."[19] The newspaper also advertised for job printing at 522 Clay Street, an address which was later to have another Spiritualist publication with woman typesetters. In mid-January 1869, the *Alta* announced that "The Women's Typographical Association propose to establish a weekly paper, to be called *The Eldorado* [sic] which shall be their organ on this coast. The paper will be conducted by women, the type set by women, and all the necessary work, except the most severe manual labor, performed by women . . . ."[20] When the first issue of *The El Dorado* appeared on 19 February 1869, it was under the aegis of The El Dorado Publishing Association. The long list of its "stock holders," printed

Fig. 16, facing page: *An example of Mrs. Richmond's use of Caslon types, then undergoing a "revival" after many years of neglect.* Author's Collection.

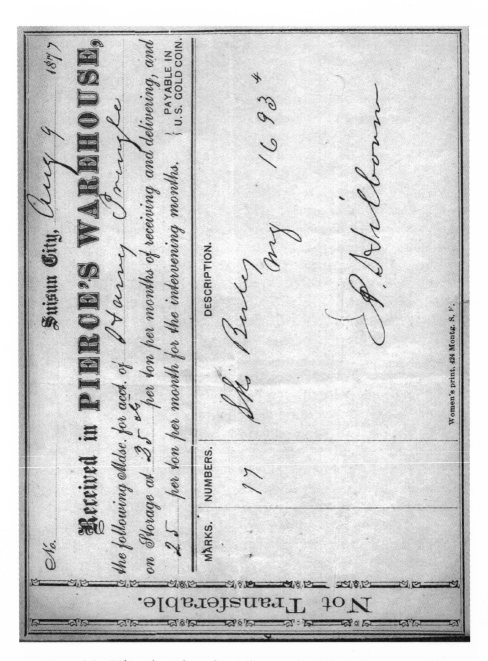

Fig. 17: *Mrs. Richmond actively sought out-of-town work and apparently was successful.*
Author's Collection.

84

below the editorial page masthead, revealed the majority were firms and individual businessmen in San Francisco, including some prominent names, such as Samuel Brannan, "capitalist" in the 1870 Census; Edward Bosqui, a leading printer; Roman & Co., booksellers; and Bradley & Rulofson, photographers.

In its salutary, the new venture was forceful: "As for the ballot box, be that not our care . . . . [*The El Dorado*] is what is called a 'woman's paper' — i.e., a paper conducted by women, and devoted to the interests of the women of the Pacific Coast. It is *not* and will *not* be an advocate of 'Woman's Rights,' as e-nunciated by certain hot-headed dames, whose zeal runs away with their discretion; but, its mission will be simply to defend and advance the social and business interests of the women of this coast, and also to afford them an organ through which their grievances may find . . . a public utterance."[21] This language certainly would appeal to the male-dominated business community that was the mainstay of the paper's support. One newspaper said the enterprise started with a capital of $20,000 and about 200 members. A reporter wrote, "By invitation . . . we visited the composing room, and saw the young ladies 'sticking type' just like professional typos."[22]

The composition in *The El Dorado* shows few of the vagaries of *Pacific Monthly*, although some of the names in the list of supporters are misspelled. An exchange commented most favorably on its typographical excellence, while noting that women were doing the work.[23] The heavier chore of press-work apparently was done by Francis & Valentine for the first three numbers, and the May to September issues were completed at John H. Carmany & Co., to judge from advertisements, passim.

Three lines at the bottom of the two, long columns of supporters said, "Publication Office — No. 517 Clay street, at the Women's Co-operative Printing Union Rooms (over the *Bulletin* office)."[24] This information must have led to misunderstandings because a disclaimer appeared in March: "We observe that a number of our exchanges run away with the idea that the paper is connected with the 'Women's Co-operative Printing Union.' This is an error. *The El Dorado* . . . has no connection with any printing office in the city. We publish our paper on our own account and furnish employment to at least twenty women, as writers, compositors, solicitors, carriers, business agents, etc."[25] To add to the confusion, the same issue contained an advertisement for the "Women's Co-operative Typographical Union [sic], Rooms of the 'El Dorado Publishing Association, 517 Clay Street, San Francisco."[26] Despite all the prominent supporters and good intentions, *The El Dorado* was short-lived.[27]

By July 1869, the WCPU was in a new location, 420-424 Montgomery Street, second floor, where the firm produced large quantities of ephemera, billheads, legal briefs and books, both paper- and hardbound, during the following eighteen years.[28] After a shaky start under Mrs. Peterson, it was now under competent new management and never faltered. The WCPU survived to become the bellwether for women in printing in San Francisco principally because of the determination of Mrs. Emily A. Pitts (A 207, A 208; see Chapter 6), editor of the *Saturday Evening Mercury* (later, *The Pioneer*), the first suffragist journal in the West. She wrote: "Some months since, through the energies of Mrs. [Agnes B.] Peterson, this business [WCPU] was established to afford employment of those women who were excluded from the other offices in this city. As a private enterprise it had been successful but still on too limited a scale to give employment to many. It is proposed to enlarge it and make it an incorporated Printing Union, conducted on the same principle as the Co-operative Union on Second St. Responsible trustees have been secured . . . . We trust our public will respond generously to this call and give our women a chance to labor on the broad platform of equality."[29]

In June 1869, Mrs. Pitts printed the formal prospectus for the stock offering and stated the case somewhat differently:

"The [Women's Co-operative Printing Union] has been in successful operation since September 1, 1868. It is unnecessary to state that women have proved their ability to execute Printing in all its branches. It is a business for which women are peculiarly adapted; and especially those inclined to intellectual pursuits. It is not more laborious than sewing and far more lucrative, the average wage paid to girls in this office being two dollars and a half per day."[30] Thus Emily Pitts's drive and determination were all-important in the incorporation of "The Women's Co-Operative Printing Union," on 16 June 1869. Its stated purpose was "To give employment to women as typesetters, and thereby enable them to earn an independent and honest living, and to conduct and carry on a general Printing business."[31]

There is an implication that Emily Pitts, and possibly others involved at this time, believed that the WCPU was a genuine co-operative. However, a co-operative modeled on the one formed in San Francisco at an earlier date would have included a payback to patrons as well as profit-sharing by the workers. No evidence has surfaced that the WCPU ever operated this way; it appears to have been strictly a stock company from the outset.[32] During its most successful years, when Mrs. Pitts no longer held sway, Lizzie G. Richmond's (A 219) astute management made the WCPU a thriving business like many another stock company in a growing San Francisco.[33] (See Chapter 7.)

At 420 Montgomery Street, Emily Pitts lived, published her newspaper and guided the WCPU. She was ever-watchful of its affairs and signed as one of the incorporators. The new legal entity was a first step in making anything resembling a "co-operative" out of the enterprise. However, stock, at ten dollars a share, was available to everyone and Emily Pitts's *Mercury* urged its purchase through a recurring advertisement.

Others signing the incorporation papers were Mrs. Mary Redding Clements, wife of an auctioneer and commission merchant; Mrs. Lucy Stuart, widow; Mrs. Martha A. Burke, housewife; Joseph W. Stow, manager of Russell & Erwin Manufacturing Co.; and J. K. S. Latham, superintendent of the banking department, Wells, Fargo & Co., whose office was nearby on the northwest corner of Montgomery and California Streets.[34] Stow was one of the prominent businessmen who supported suffrage and other woman's causes. His wife, Marietta (A 262), later became editor of the *Woman's Herald of Industry* and heavily involved in printing and feminist activities (see Chapter 8).

Following incorporation, Mrs. Lizzie G. Richmond, the new superintendent, gradually gained full control of the business as Emily Pitts put her energies into the suffrage cause and *The Pioneer*. Despite the fact that there was a total of fifty-four job printing-offices in the city by 1870, the WCPU succeeded under Mrs. Richmond's leadership and thus helped women to become a permanent, albeit small, part of the printing industry in San Francisco.[35]

Chapter 6   Printing, Publishing and Polemics

Emily Pitts Stevens rose to prominence when she converted the *Sunday Evening Mercury*, which styled itself "a Journal of Romance and Literature," into the first voice for woman suffrage in the West.[1] She hired women to set type for her newspaper; helped to promote successfully one major printing business, the Women's Co-operative Printing Union, so more women might have work as typesetters; and in 1872 founded the Woman's Publishing Co. for the same purpose.

Yet, Emily Pitts Stevens remains a woman of mystery. An account of her life, published in an 1893 directory of women, contains some verifiable facts, garbles others and omits any reference to her life before she came to San Francisco.[2] As a final irony, her name is misspelled on the family gravestone.[3] There is no information on her former life in the East except the year of her birth, 1841, and the place, "New York," city or state not specified.[4] However, scattered references suggest she came from at least a middle-class background, had learned to set type, apparently knew Susan B. Anthony and Elizabeth Cady Stanton, the suffragist leaders, and was a reformer.

In 1871, Miss Anthony described Ms. Pitts in a manner that delineated her character, declaring her "a nervously organized, pleasing little woman, with dark eyes, curly hair, refined manners and features, her every word and movement indicating culture and good breeding." As San Franciscans would learn, "She speaks and writes with force and point," and because of this ability, though not her arguments, "was thoroughly respected for her earnest devotion to her principles and the ability with which she advocates them."[5]

She arrived in San Francisco in 1865, a mature woman of twenty-four, using the name Miss Emily A. Pitts, to work as a teacher. Five years later, she recalled: ". . . [W]e [Pitts] came to San Francisco from New York, in company with Miss [Florence] James, daughter of the celebrated novelist, G[eorge] P [ayne] R[ainsford] James,[6] for the purpose of establishing a Female Seminary. For two years we [Pitts and James] were in company. For two years we [Pitts and James] held, as teacher, an important and responsible position in our establishment, the Miehl Institute."[7] Then, in 1867, at $50 a month, she "taught and built up under the superintendency of Mr. [John C.] Pelton, the public evening school for girls. Our health failing us under the labor of teaching day and evening, we resigned our evening charge."[8]

Once she had lightened her teaching load, Emily Pitts turned to journalism, for she wished to aid women through increased job opportunity and political equality. "By dint of untiring industry," she said in an 1870 autobiographical sketch, "we purchased the old *Sunday Mercury* and at once gave employment and pay to needy women, and encouraged others to learn the printer's art."[9] Additionally, she reached more people through a broader-based crusade, and "have done what [we] could to awaken, all over the Pacific Coast, the justice for women's claims for the ballot," and the "necessity" for women to vote "at once."[10] Her name appeared in the masthead as "Mrs." Emily A. Pitts.

Fig. 18: *Mrs. Emily Pitts Stevens, whose name was misspelled on this San Francisco Call likeness, as well as on her tombstone.* Courtesy The Bancroft Library.

She had a partner, writer - printer Frank S. Wickes (C 11), in the acquisition of her newspaper. While Mrs. Pitts directed editorial policy, Wickes supervised production and remained in the background. Upon purchase, the new editor proclaimed, through her Salutatory on 24 January 1869: "The wrongs of woman, the many abuses she has suffered, have at last [been] aroused from their lethargic sleep. . . . We shall claim for her the right of suffrage — believing by this she will gain the position for which God intended her — equality with man."[11]

The first front-page "Woman Suffrage" story appeared on 25 April, next to the usual serial romance, and continued there in the next edition. Regardless of the paper's makeup, Mrs. Pitts hammered away at the themes outlined in the Salutatory, starting with her second issue. First came "Woman as a Revolutionist," a theme definitely not heard before in San Francisco.[12] She then declared, "Woman's Suffrage is not a struggle in which men are to be the losers and women the gainers. It is a struggle in which both are to gain and lose alike."[13] Finally, she summarized her view succinctly in an en-

91

during sentence, "The question of the age is one — exclusively woman's rights."[14]

Under the heading "Eight Hours for Working-girls," she wrote a romanticized description of how young females in San Francisco —including typesetters — started and ended their workdays and tied it directly to the new movement towards an eight-hour day: "Early in the morning, before seven o'clock, we have observed many young, industrious, and pure-minded girls hurrying along the street" to work. "In the evening late, they are seen trudging homeward, not tripping so lightly as in the morning . . . but, notwithstanding the vigor of youth, they are tired — very tired. Towards night . . . the book leaves have folded more reluctantly, and the foundry type[s] have begun to abrade their tender fingers more and more, as the shadows slanted across the way. . . . We have thought that while the young working men of San Francisco are for themselves establishing an eight-hour reform, they should exert their influence to procure the application of the same rule in those establishments where young women are employed."[15]

Emily Pitts rejected a claim that "when women were engaged as compositors, they were not paid equally with the men employed in the same offices." She rejoined, "We emphatically deny that statement as far as the *Mercury* is concerned; those employed by us are paid the same as the male compositors. We would scorn to advocate a principle and not live up to it. We believe in equal rights, and 'we practice what we preach.' "[16]

In May, the partners changed the day of publication, the name to *Saturday Evening Mercury* and announced, "Our circulation has largely increased . . ." as more and more people evidently were welcoming this pioneering voice for woman's rights.[17]

Regardless of other causes, Mrs. Pitts never abandoned woman typesetters. The struggling Women's Co-operative Printing Union was always a close concern and both were tenants at 420–424 Montgomery Street. "It is unnecessary to state that women have proved their ability to execute Printing in all its branches. It is a business for which women are peculiarly adapted; and especially those inclined to intellectual pursuits. . . the average wage paid to girls in this [WCPU] office being two dollars and a half per day." The remainder of this article is one of many continuing pleas for support of the corporation about to be formed by stock purchase.[18] Later, Emily Pitts explained: "The object of selling shares is to obtain capital to purchase more material, in order that more women may be employed, and more young girls can learn type-setting. Constant applications are made for positions, which must be rejected, owing, not to want of work, but to want of type." She concluded "that as far as their limited material will permit . . . they can compete with any office in the city."[19]

Mrs. Pitts continued to press for acceptance. The WCPU had a display at the Industrial Fair, and she prompted "good friends" to "examine these specimens" and "leave orders." She wrote that since "women have but a poor chance to compete with the work of men, to give them that chance."[20] "What we must do is to develop a public sentiment which makes labor honorable in wom[a]n as well as man."[21] Following up in late 1869, Emily Pitts wrote *The Revolution* on behalf of the WCPU: "We are in want of three, good, reliable women, skillful typesetters, who would be willing to come to San Francisco. We will give them steady employment, at the rates we are now paying, fifty cents per thousand [ems]. They can become shareholders in the [Women's Co-operative Printing] Union which has been established solely for the benefit of the working woman."[22]

On 13 November 1869, Emily Pitts came into full control of her newspaper and changed its name. "We are now the unchallenged proprietor of *The Pioneer*. We have paid the last dollar on it. . . . *Pioneer* is a name that more nearly covers our thought and tells the nature of our object and ambition. A pi-

oneer signifies one who goes before with pick in hand to pre-
pare the way for others to follow. . . . She also noted: "In addition to our editorial labors we do our own soliciting. We make
our own collections. We assist in mailing each edition of our
paper and *devote the proceeds of teaching* to its advancement [em-
phasis supplied]."[23] Her revelation that she was continuing to
teach and thus earn money for her newspaper might explain
its survival — and hers. There is no known record of this later
teaching activity.

Coping with a new publishing venture and promoting
the WCPU did not deter Emily Pitts from spreading her energies still further. She soon turned her attention to suffrage
and became a driving force in the founding of the California
Woman Suffrage Association on 28 January 1870.[24] *The Pioneer*
constantly pulled suffrage news from across the country,
sometimes printing entire speeches by Eastern leaders, such as
Elizabeth Cady Stanton and Susan B. Anthony. Though
schisms divided the woman's movement, Mrs. Pitts remained
aligned with Mrs. Stanton, Miss Anthony and their followers.

Despite the growing demands on her time, *The Pioneer* did
not suffer, as the *Pacific Monthly* had under Lisle Lester, even
though over the years Emily Pitts apparently wrote less and
less. Frank Wickes probably was of inestimable assistance because the newspaper continued all the while to promote her
interests and causes.

A short time later, Emily Pitts's personal life took anoth-
er turn. She both lived and worked at 420–424 Montgomery
Street, the building into which the WCPU had moved in July
1869.[25] Other tenants included Samuel Brannan, a noted San
Francisco pioneer, and one of his employees was Augustus K.
Stevens, a 25-year-old accountant and Massachusetts native.[26]
Stevens appears in the 1867–1870 Directories as also living at

Fig. 19, facing page: *Advertisement in* Nevada County Directory for 1871–1872. Courtesy The Bancroft Library.

The Pioneer.

LIBERTY, JUSTICE, FRATERNITY.

Devoted to the Promotion of Human Rights.

This Office was the first to throw open its doors to women, in spite of the arbitrary and unjust laws of the Printers' Union, which proscribed all women from a Printing Office

NOT ON ACCOUNT OF ABILITY BUT SEX!

What is the Result?

Many of the largest and best Printing establishments on the Pacific Coast have thrown open their doors to all comers who can do good work.

T E R M S :

One Copy one year $3 00
One Copy six months 2 00
Five Copies one year 12 50

Send Money by Post Office Order or Express.

EMILY PITTS STEVENS,

Editor and Proprietor.

Office—420 Montgomery St., San Francisco.

the Montgomery St. address. An attachment evidently evolved from this proximity because the 1871 Directory lists "Mrs. Emily Pitts Stevens" as living at 538 Greenwich St. with Augustus K. Stevens.[27] A new name, "Emily Pitts Stevens" ("Mrs." and "Emily A"were added later) first appeared in the masthead of *The Pioneer* on 8 January 1870 but there is no traceable information about this marriage.[28] Mrs. Abigail J. Duniway, the pioneer suffragist of the Northwest, following a visit to the Stevens's home on Greenwich St., found Mr. Stevens "a genial, happy, whole-souled gentleman, with love-light in his eyes, and a pardonable pride of his gifted wife in his heart . . . ."[29]

By 1870, Emily Pitts Stevens had become a power in woman suffrage. She was not only a principal in the Woman Suffrage Association, but also active with a petition to the California legislature seeking a constitutional amendment to grant woman suffrage. This petition included 3,300 names, many of them men. A random check of the unmarried women who signed from San Francisco shows a preponderance of schoolteachers, their assistants, women in the needle trades and only one typesetter, "Miss M. Smith." Miss Margaret (Maggie) Smith (A 245) is listed as working as a compositor at the *Saturday Evening Mercury* in the 1870 Directory and continued to work in the trade for many years.

Mrs. Pitts Stevens and her husband were signatories, of course, as well as Frank Wickes. Many others who were to remain in the forefront of the suffrage movement also signed, such as Mrs. Elizabeth Schenck, the former editor of *Pacific Monthly*; Mrs. Eunice Sleeper, a longtime suffragist; Mrs. Carrie Young, editor of the *Women's Pacific Coast Journal*; and John A. Collins, a lawyer, Spiritualist, former state legislator and political gadfly, who was a leading male suffragist.[30] I. N. Choynski, the bookseller and publisher, and other businessmen also signed this trailblazing petition.[31]

The Pioneer thanked the railroads "for their kindness in charging but half fare to the women delegates who wished to travel to Sacramento for the presentation of their petition before the Assembly."[32] However, the Legislature took no action. The lack of success did not discourage Emily Pitts Stevens and her friends from continuing the political battle vigorously.

An 1870 newspaper strike (see Chapter 7) finally ended Union dominance over the policy of non-employment of women in these better-paying establishments. Said *The Pioneer:* "*The Morning Call* office has five women at the case in its composing room and speaks highly of both their skill and industry. The *Call* and *Bulletin* are now in a condition to afford women printers a helping hand. The proprietors of both these journals have notified the public that they are now free to employ whom they please. Had there been one hundred women, thoroughly skilled in the art of type-setting at the time of the strike, they could all have secured remunerative positions and those male printers too selfish or too proud to work by their side would have been permitted to emigrate to the Country, Canada or China."[33]

However, Union records show that eighteen years later the *Call* had but four women on its roster of eighty-one employees while the *Bulletin* had only four women among its forty-four workers. The percentage was even worse in the book-and-job offices which showed five women within the total of well over three hundred.[34] Only the woman-controlled, non-Union offices consistently had a higher ratio of women to men employees.

Mrs. Pitts Stevens's follow-up comments were even more pointed: "The Typographical Union was a narrow, selfish and despotic body" which made its defeat "a victory for women . . . on the entire Pacific Coast, of no light consideration. . . . If women would throw aside the foolish idea that it is lady-like to teach school but vulgar to set type, or perform what is ordinarily called work . . . they would soon be in a condition to command respect and influence… ."[35]

The Pioneer often gave news about Emily Pitts Stevens and

BACKWARD GLIMPSES

GIVEN TO THE WORLD

BY JOHN BUNYAN,

THROUGH THE INSPIRATION OF

SARAH A. RAMSDELL.

———————

SAN FRANCISCO :

Woman's Publishing Company, Printers, 511 Sacramento Street

1873.

Fig. 20: *Spiritualist title printed by woman-run printing-office.* Courtesy
The Bancroft Library.

itself from exchanges, an interesting example being a *Vallejo Chronicle* story about a Woman's Suffrage Association meeting held at Solano City (16 miles southwest of Sacramento): "On taking the stand, Mrs. Stevens said that she was not a public speaker, and had never before appeared before an audience in that capacity on but one or two occasions. Nevertheless, she succeeded in holding the attention of the audience for a considerable time, speaking with a remarkable fluency and considerable force. . . . Her earnestness of manner — by no means unladylike — was a subject of general remark."[36] Another exchange added,". . . the lecture was replete with grace of manner, harmony of voice and with that eloquence of earnestness that always produces conviction, and filled with cogent arguments why women should be enfranchised."[37]

Mrs. Pitts Stevens spoke twice at the first Pacific Coast Woman Suffrage Convention in San Francisco, 17–18 May 1871. During an evening session, which was so crowded that women had to stand, she took the anti-feminist *Chronicle* to task by denying that she had said Laura Fair, an accused murderess in a sensational local case, was a courtesan.[38] "I said nothing of the kind. What I did say was, 'she is as good as those male courtesans.' (Mrs. Pitts Stevens threw a copy of the paper on the floor as she spoke)."[39] At adjournment, it was announced that Susan B. Anthony would soon visit San Francisco and plans should be made.[40]

Ms. Anthony and Mrs. Elizabeth Cady Stanton arrived in July 1871, and the *Chronicle* greeted them with headlines which proclaimed: "The Female Agitators; The Great Woman Suffrage Champions Here."[41] They had come from Utah escorted by Mrs. Pitts Stevens and were the guests of Leland Stanford at the Grand Hotel.[42]

Mrs. Stanton first spoke on "The New Republic" in Platt's Hall "before an audience that crowded every part of the building" after having been introduced by Mrs. Pitts Stevens. Her talk "excited the greatest interest," the *Call* reported, "while every now and then general and sometimes even re-peated applause" greeted her words on behalf of woman's suf-frage. She also commented on the Fair trial, saying, "Whatever might be the merits of the case which had just agitated society in San Francisco, no one had any right to hang a woman until she had been tried by a jury of her peers. (Applause and a few hisses.)"[43] Emily Pitts Stevens and the visiting suffragists at-tended the Fair trial and said "she had been singled out of the crowd in attendance and fined for a violation of decorum by applauding."[44] (Mrs. Stanton, Ms. Anthony and Mrs. Pitts Ste-vens also visited the defendant in her jail cell.)

The following evening, "Mrs. Pitts Stevens came upon the platform and placed upon the speaker's table a handsome bouquet of roses." Susan B. Anthony then spoke on "The Pow-er of the Ballot," after having been introduced by Mrs. Eliza-beth Schenck. She also commented, at greater length, and in a more polemical style, on the Fair case, saying at one point, "If all men protected all women there would be no Mrs. Fairs." She concluded her lecture "with an earnest appeal to all to support Mrs. Pitts Stevens in her proprietorship of *The Pioneer*."[45]

One newspaper commented harshly: "Many left the hall disappointed. . . . Any favorable impression made by Mrs. Stanton . . . Miss Anthony succeeded in removing. . . . Mrs. Stanton should be permitted to make this campaign alone."[46] Mrs. Stanton returned to the podium to speak on "Woman Suffrage and Free Love" before a large, general audience. The *Call* reported: "Mrs. Stanton was preceded, as usual, upon the platform by a little advance guard of the strong-minded, among whom we noticed Mesdames Schenck, Sleeper, Snow[47] and other veterans. . . . Mrs. Stanton appeared at last, accom-panied by her chief aid[e], Mrs. Pitts Stevens . . . ." Suffrage was the core of her long discourse but at one point she de-clared: "The great bugbear was the question of Free Love. She had been accused of holding free love doctrines but had really had no time to love more than one. (Laughter.) . . . All this cry about free love is arrogant hypocrisy. There is not a man who writes a line about it who does not know he is a hypocrite. That is the truth."[48]

Emily Pitts Stevens herself had been engaged in the Free Love dialogue as early as 1869 when the *Bulletin* had taken a stand against the idea that a female should "pop the question" when it came to marriage. She replied with strong, unequivocal words: "Free love, as we understand the term, means a free and untrammeled intercourse between the sexes, and that any public sentiment or legal enactments that impose restrictions on human passions, or limits their actions, are arbitrary and oppressive."[49] The Free Love issue was to trouble her for a long time.

A new political assault on Sacramento took place in March 1872. "The California Woman Suffrage Association has delegated Mrs. Emily Pitts Stevens . . . and others to address the members of the Legislature," an Oakland newspaper announced, "on the 'Fundamental Principles of Republican Government.' They will occupy the Assembly chamber for that purpose next Wednesday evening [20 March]"[50] before "A Joint Committee of Senate and Assembly, to whom was referred certain petitions for an Act authorizing women to vote . . . . The floor and gallery of the chamber were thronged and great numbers were unable to obtain seats." Unfortunately, the meeting was a debacle and "increased so in rude boisterousness that the Chairman, justly incensed, adjourned the meeting."[51]

The *Chronicle's* brief report of events had its usual tone of derision and took aim directly at Mrs. Pitts Stevens: "Then came 'our' Emily Pitts [sic], and she lectured the crowd, who received her remarks with good humored and frequent — too frequent — applause. Another old [!] lady followed. Her voice was cracked, and she lifted it very high, and people laughed immoderately. Pitts looked daggers, and the applause increased. [Chairman] Finney looked troubled, and the trouble

Hon W. T. Sexton

from Rhodes

NO.

IN THE

SUPREME COURT

OF THE

STATE OF CALIFORNIA,

A. McDERMOTT, et al.,

Respondent,

v.

H. K. MITCHELL,

Appellant.

Petition for Rehearing on part of Respondents.

W. H. RHODES,

Attorney for Respondent.

SAN FRANCISCO :

Woman's Publishing Co., 511 Sacramento St.

1874.

increased. Finally, the applause went in advance of the speech and the Chairman, becoming disgusted, declared the seance [!] adjourned. The audience enjoyed the performance; it was better, they all declared, than a circus. What its effect will be on the Legislature we are unable to say. It never was intended to do anything on the subject, and the views expressed last evening will not improve the prospects of the Female Suffragists. Let us crow."[52]

Despite the increasingly time-consuming and emotional stress of her political activities, printing was never far from Mrs. Pitts Stevens's mind. She next signed an application for incorporating a new printing venture, filed at the City Clerk's office in San Francisco on 13 April 1872. It declared: "The Woman's Pacific Coast Publishing Co. is formed to carry on the business of book, job and all kinds of printing, bookbinding and engraving, to publish books, magazines, newspapers and deal in printing paper and stationery, books and periodicals.

"Capital stock shall be Twenty five Thousand Dollars divided into Twenty five Hundred (2,500) shares of the par value of Ten dollars ($10.00) each. ... The number of its Trustees shall be three and that the Names of those who shall be Trustees, and manage its affairs during the first Three Months, and until their successors are elected, are Frank Sleeper, A. K. Stevens and Frank Wickes." The document was signed by all three as well as Emily Pitts Stevens and Eunice S. Sleeper, her suffragist colleague, whose husband, a farmer according to the 1872 Directory, had also signed the 1870 Petition.[53]

A recurring, column-length advertisement in *The Pioneer* later amplified the cold details of the incorporation papers with informative rhetoric: "Equal Pay for Equal Work," read the heading. "Assist Women in Honorable Employment. Woman's Publishing Company, Printers, 511 Sacramento Street, San Francisco. This printing establishment has been

Fig. 21, facing page: *This represents the general format for legal briefs of the time. Rhodes was also a well-known Spiritualist.* Author's Collection.

founded for the purpose of opening up a school of instruction for girls and women, who are dependent and wish to learn the art of type-setting and printing as a means of livelihood. A first-class job office has been purchased, with type and presses unsurpassed. Experienced and expert printers are engaged, capable of executing every class and variety of book and job printing, in superior style. CITIZENS! Will you aid this enterprise by giving us a portion of your printing, which will be guaranteed equal with that of any other house, and at fair rates[?] Lawyers' briefs, books, cards, billheads, statements, circulars, posters, etc. We wish to be able to give employment to at least FIFTY women and girls. Large numbers of applicants are necessarily refused, as present facilities are not equal to the accommodation of all. Lend us a helping hand, that we may increase our sphere of usefulness."[54] At the foot of the column the ad read, "Emily Pitts Stevens, President and Superintendent," which explained the course of events after incorporation.

There is a possibility that Mrs. Pitts Stevens acquired existing equipment because her new address, 511 Sacramento St., housed two printing firms of record in the immediate past: Walters, Newhall & Co., 1869–1872, and Smyth & Shoaff, 1869–1872.[55] At once she began to set type and accomplish the presswork for The Pioneer at the new address, which duly appeared in her masthead as her editorial office.

On the political front, a troublesome matter was the candidacy of George F. Train, a suffragist, for president in 1872. He first visited San Francisco in 1869 to give a series of lectures and The Pioneer described him as "that erratic genius and bundle of incongruities ... one of the most extraordinary of human riddles."[56] Notwithstanding, Mrs. Pitts Stevens acclaimed his return: "Our next president is here! George Francis Train has advocated Woman Suffrage for the last twenty years. We want a Woman Suffrage President."[57] She then placed an advertisement beneath her editorial masthead which ran until the election in November, reading in part: "The Arrival of the Comet!

Veni! Vici! We feel it! We believe it! It is in the air! . . . George Francis Train as our next president! We have discovered that it means We, the People! He represents the spirit of the Age. *Vox Populi! Vox Traini!*"[58] San Franciscans, including many of her supporters, saw her extravagant support of Train as capricious and ill-advised and they were correct within the context of the times.

The *Chronicle* renewed the political fray in June 1872 by including Emily Pitts Stevens's name in a long, front-page story headed, "A Nest of Free Lovers," an exposé of the Radical Club where a reporter had been attending meetings.[59] This was soon followed by patronizing coverage of the second Pacific Woman Suffrage Convention in mid-June under headlines reading, "Cluck, Cluck, Cluck," and "The Cackling Chorus." Emily Pitts Stevens told the gathering: "I suppose every man and woman knows what this Convention is called for. It certainly is not for the purpose of discussing free love . . . nor for anything but the political enfranchisement of women."[60] Unfortunately, the question continued to be attached to her persona whether she willed it or not.

This Woman Suffrage Association convention met for four days. Emily Pitts Stevens was not only named president but also was chosen to head the newly organized Pacific Slope Suffrage Association, created to include California, Oregon, Nevada and Washington Territory suffragists in one organization.[61] An *Examiner* reporter reverted to style when he said Mrs. Pitts Stevens "spoke in her usual forceful and logical manner."[62] Otherwise, this meeting concluded without major problems.

However, a few days later, an anti-feminist lecture was given by the same person who had followed the San Francisco women before the legislative committee in Sacramento. The event became a matter of hilarity for the press but left Emily Pitts Stevens scrambling to protect her reputation. The affair began with a somewhat ambiguous story in the *Examiner* on 24

June: "Mrs. J. B. Frost is announced to deliver a lecture this evening at Platt's Hall. Mrs. Frost is down on the bills as the anti-railroad monopoly, anti-Goat Island grab,[63] anti-Woman Suffrage, anti-Spiritualism and anti-Free Love orator."[64] The *Call* said Mrs. Frost was announced to lecture "Anti-Railroad" and that only forty people showed, because they "prefer to receive their anti-railroad information through some better recognized source . . . ." Its reporter found "a small but picked cohort of woman suffragists" in the front seats, including Mrs. Emily Pitts Stevens, Mrs. M. O. Loomis, Mrs. M. L. Olmstead and Mrs. C. M. Churchill who were present to defend [railroad magnate] Leland Stanford with their applause if necessary."[65]

The *Call* also gave the most detailed account of the evening, starting with the change of subject. Mrs. Frost said "she had learned too late that her 'agent' had advertised her to speak on the Railroad Question and . . . she had not yet had time to study up on the subject. She would therefore give . . . her lecture against woman suffrage." Her remarks "pleased not the select cohort of the strong-minded. They applauded at the wrong places . . . they laughed sarcastically, and they passed exceedingly critical remarks in very audible stage whispers. A male challenged the women, saying, 'I move that these ladies behave themselves.' 'Second that motion! Second that motion!' cried Assemblyman [David] Meeker, jumping to his feet.

"Mrs. Pitts Stevens, pale with passion, rose to her feet and replied, 'We have a right to applaud if we please.'

"Assemblyman Meeker: 'I intend to make you behave yourself Mrs. Stevens. You are behaving disgracefully.' [Applause, hisses, and as much confusion as an audience of forty could make.]

"Mrs. Stevens: 'You will find it difficult to do anything that we don't like.' [Confusion worse confounded.]"

A "war of words" ensued and finally the lecturer resumed but added fuel to the fire by declaring "that by insuring his life in favor of his wife a man puts his life in her hands, and she did not see what the Woman Suffragists wanted to do unless it was to kill their husbands. [Hisses from the Woman Suffrage cohort.]" Matters rapidly went downhill from there with Mrs. Pitts Stevens being shouted down. Mrs. Frost went on to say that there were 20,000 prostitutes in New York, adding, "That this is the class of women that would elbow their way to the polls. Mrs. Loomis: 'I suppose it is the same class of men who do the voting, then.' [Hisses and applause.]" Finally, everyone settled down again.

Following the meeting, the *Call* reported a further problem: "Nevertheless, as the audience was dispersing, and Mr. Meeker was walking down the aisle on his way to the door, Mrs. Stevens literally went for him. Attended by her friends she reached his side and demanded an apology. He refused to give it and retorted that she had misbehaved herself. Bandying words of no very amiable character, the party reached the door leading to Montgomery Street when it was noticed that Mrs. Pitts Stevens had a derringer in her hand raised to about the height of her chest, and in close proximity to Mr. Meeker's. A bystander [another paper said it was the *Call* reporter[66]] caught hold of it to take it away from her, but she said, 'I will put it in my pocket,' and she did so. Mrs. Stevens walked on . . . . We understand that the pistol did not belong to Mrs. Stevens, but had been given to her just previously by Mrs. Churchill. The whole affair was very unfortunate. The ladies allowed their zeal for the Cause to make them forget the rules of politeness and good order, and Mr. Meeker forgot that he was talking in public to a woman."[67]

Emily Pitts Stevens immediately wrote to all the newspapers: "I have been most unjustly accused by the Press. I never pointed a pistol at Assemblyman Meeker . . . . Being a member of the Peace Society, I am totally opposed to the use of firearms." She concluded, "I am persecuted by opponents of the Woman's Suffrage cause . . . ."[68] *The Pioneer* also denounced "The Falsehood about Emily Pitts Stevens and Her Pistol" with a strong denial but the net effect of the whole episode was to diminish her standing within her own constituency.

The following January, Mrs. Pitts Stevens pontificated: "With the New Year we shall enunciate new ideas. . . . *The Pioneer* for 1873 will take large, generous and Catholic views of human events. While the enfranchisement of women will be the corner stone of its faith ... it will in no sense be governed by the 'one idea' theory."[69] Despite this affirmation, California suffragist politics came to a boil the following year. Emily Pitts Stevens was at the center of events starting with the annual Pacific Slope Convention, now under the leadership of John A. Collins. He was a man of varied political experience in the East, Nevada and California, had been a printer, later became a lawyer and moved to San Francisco in 1865. Collins was widely known as a political radical, a "conservative Spiritualist," as well as a supporter of woman suffrage. By the time of the 1873 gathering, he had antagonized many women over a period of time by his dictatorial ways. One newspaper called him "Field Marshal Collins."[70]

Serious problems began with the appearance of a New Yorker, Mrs. Anna Kimball, an ardent supporter of Victoria Woodhull, Free Love, suffrage and Spiritualism. The *Chronicle* said: "The convention later in the evening adjourned in a row, and now it is said that the Kimball party are going to organize a convention of their own, to advocate Woodhullism [i.e., Free Love]. This is probably the end of the suffrage movement in this State for the present."[71] The next day, "After a parliamentary maneuver [i.e., a "fast gavel"] to adjourn by John A. Collins, which caught Mrs. Pitts Stevens and her followers off-guard, Col. Collins acquired the bank funds, records and all else of the Woman's Suffrage Association [organized 28 January 1870] and left the hall."[72]

Mrs. Pitts Stevens and her supporters decided to form a new organization the next day, specifically excluding John A. Collins and Mrs. Amanda Slocum. Mrs. Pitts Stevens "arose to a personal explanation," said the *Chronicle*. "With tears in her eyes, she disclaimed any belief in free-love heresies and bitterly denounced the *Chronicle* reporter for classifying her with people who are open in their advocacy of these damnable doctrines. After that she sat down and had a good cry, and then came to the *Chronicle* man and humbly apologized for having abused him in so virulent and unseemly a manner." Graciously, "The *Chronicle* man accepted the apology and, taking the article, pointed out to the well-meaning but impulsive lady that in no way was her name connected with the doctrine of Free Love and that in putting the cap on her own head she had acted imprudently — in fact, had gone off half-cocked. After that all was lovely with Mrs. Pitts Stevens . . . . The Woman's Suffrage Society was then formed with Mrs. Stevens as Corresponding Secretary."[73] This action, of course, duplicated a society already in existence and added to the divisiveness and confusion in the woman's movement in California.

Matters simmered for almost two weeks before a letter, more than fifty column-inches in length, ran in the *Chronicle* on 27 April 1873, from officers of the California Woman Suffrage Association. Signers included Mrs. Eunice S. Sleeper and Mrs. Mary G. Snow, erstwhile close associates of Emily Pitts Stevens, as well as Mary J. Collins, wife of John A. Collins and a longtime suffragist in her own right. They recalled with shame and sorrow Mrs. Pitts Stevens's unfortunate, threatening and revolutionary harangues from the disruption in the Legislative chambers in March 1872 to the present. In regret, they concluded, "Mrs. Stevens is indebted to her own acts and the unwise counsel of her pretended friends."[74]

All in all, the letter was an indictment of Emily Pitts Stevens personally, including a rehashing of the Free Love charges. Interestingly, it revealed that "Mrs. Stevens for the first time in nearly three years, was a prompt and constant attendant at business suffrage meetings," hinting that she may have been a "no-show" in the early years. The signers also revealed how Emily Pitts Stevens saw the future, quoting her that she "was done with suffrage organizations, and hoped they would go down." In the future, Mrs. Pitts Stevens had announced, "her labors for women" would be "confined to her own [printing] business, and be of practical character."[75]

The undaunted Emily Pitts Stevens had still one more round to go in the organizational battles. Over the imprint of her Woman's Publishing Co., she issued a flyer calling for a 25th anniversary celebration "to commemorate the First Woman-Suffrage Meeting Ever Held in the United States." The speakers announced included Laura DeForce Gordon and Mrs. Pitts Stevens. The *Alta* proclaimed, "The 'Bolters' Before the Public in a New Convention" and "The Cause of the New Faction Apparently Triumphs."[76] However, the newspaper then ridiculed them for "Fighting windmills." The occasion "was selected as the proper time for the inauguration of the new Convention, out of which the wily wire-puller Collins has been ejected, or rather, which presents to him the cold shoulder" while "Mrs. Stevens read a few verses supposed to be appropriate to the occasion." Finally, Laura DeForce Gordon was upbraided for saying the press was not present — "our reporter was there and took it all meekly like a lion in a den of lambs."[77]

This gathering proved to be the swan song of Emily Pitts Stevens in the mainstream suffrage organizations. No more polemical editorials appeared in her paper, either. In October 1873, Mrs. C. C. Calhoun, the veteran suffragist, "took proprietary interest of *The Pioneer*" and declared, "We cannot forego the duty we owe our predecessor in the recognition of past services upon her part; services that should not be lightly remembered but should be kept in grateful recollection by the readers of *The Pioneer*."[78] Coincident with the takeover by Mrs. Calhoun, a new name, Emily Robins (A 221), also appeared in the recurring advertisements for the Woman's Publishing Company. She was listed only as Superintendent. Probably

WOMAN SUFFRAGE!

25ᵀᴴ ANNIVERSARY

OF THE

FIRST MEETING HELD IN THE UNITED STATES

—TO—

Advance Equal Rights.

All Woman-Suffragists are cordially invited to meet at

PACIFIC HALL,

San Francisco,

Tuesday, May 6th, at 10 A. M.

To take part in a Meeting to commemorate the

FIRST WOMAN-SUFFRAGE MEETING

EVER HELD IN THE UNITED STATES.

Friends of Suffrage are Expected from the Country.

THE FOLLOWING SPEAKERS WILL BE PRESENT:

Mrs. Laura De Force Gordon, Mrs. L. Mathews, Mrs. Roberts, Mrs. Benedict, Mrs. Damon, Mrs. Stevens and others.

Woman's Publishing Co., Printers, 511 Sacramento Street.

Fig. 22: *This rare 1873 handbill was printed by Emily Pitts Stevens's firm. The meeting marked a high point in her San Francisco political activity.* Author's Collection.

Emily Pitts Stevens, who had been president since its founding, retained her control of the corporation. The business remained active at 511 Sacramento Sreet until it was purchased in 1875 by Mrs. Amanda M. Slocum (A 239, A 268) when she and her husband gave up their Spiritualist publication, *Common Sense*, terminated because of lack of support (see Chapter 8). Thus, in a great irony, one of Mrs. Pitts Stevens's opponents in the woman's movement took control of a printing business that was founded on her strong desire to furnish training and meaningful employment for "needy" women in the community.

The disappearance of Emily Pitts Stevens from woman suffrage organizations and the sale of her newspaper by no means ended her activities, either political or journalistic. In May of 1874, she was a delegate from San Francisco to the State Temperance Alliance for California which later became part of the newly organized Women's Christian Temperance Union. (Interestingly, a fellow-delegate was Mrs. David Meeker, wife of one of her recent adversaries.) She was elected "Grand Conductor of the California Grand Lodge of the Sons of Temperance."[79] From 1875, she listed herself as journalist in the San Francisco Directory and by 1880 as a "Temperance Lecturer." A history of the movement names her as one of the "Organizers and Field Workers," and she often served as a delegate to various meetings.[80]

An 1891 news story listed some of the activities which benefited from her considerable energies after she left the suffragists. She "made possible the Women's Pacific Coast Association of Writers. . . . was the popular lecturer of the Good Templars and has been Grand Lecturer for the State for the W.C.T.U., and is now State Superintendent of Foreign Work for that organization. For a time she most successfully conducted a sewing-school for poor girls at the Silver Star House." The article appeared on the eve of her leaving for Boston to attend the World and National Convention of the W.C.T.U.[81]

A few years later, she again surfaced in a familiar context, a meeting of the Equal Rights League. Here she said "that in nearly two-thirds of the States of the Union the age of consent is greater than in this State — in a very few it was below 16. The passage of such a law would . . . be a credit to the State and the Legislature." This article was also accompanied by a likeness of Mrs. Pitts Stevens.[82]

After having lived at several addresses through the years, in 1881 Mr. Stevens purchased a lot and home at 1907 Bush Street. There were two bedrooms, front and rear parlors as well as a sitting room. The decorations included chromolithographs and steel engravings while the dining room had a large and tasteful assortment of dishes and silverware suitable for entertaining.[83] Mr. & Mrs. Stevens were listed for many years in the *San Francisco Blue Book*, "at home Thursdays at 2 p.m." for social visiting. This kind of gracious living was probably close to Mrs. Pitts Stevens's earlier experience in the East.

Gradually, Emily Pitts Stevens's name disappeared from the newspapers. Apparently she lived quietly in her comfortable surroundings. She suffered a cerebral hemorrhage 10 September 1906 and died three days later, aged 65. There were no lengthy obituaries despite her long career in the city.[84] After the facts of her death, the *Chronicle* noted: "Friends and acquaintances are respectfully invited to attend the funeral services, Saturday, September 15, at 1 o'clock at her late residence, 1907 Bush Street. Internment, Mt. Olivet Cemetery by funeral [trolley] car from Thirtieth St. and San Jose Avenue."[85]

She lies in Mt. Olivet (Colma, San Mateo County) beneath a headstone which bears no dates for her and with Pitts misspelled, errors undoubtedly made by the conservator who ordered the stone for the family grave following Augustus Stevens's death on 4 June 1912. Her notable achievements, the establishment of the first suffrage journal in the West and her success in helping women to obtain training and permanent work as typesetters in San Francisco, have too long gone unheralded.

Fig. 23: *420–424 Montgomery St. with Women's Co-operative Printing Union sign clearly visible in upper left. Horse car is in motion on the tracks which head south to the original Palace Hotel. California St. is the cross street. The Wells, Fargo History Museum is in this space today with a new building.* Author's Collection.

Chapter 7   Lizzie G. Richmond Creates a Success

Of all the woman-controlled printing establishments, the Women's Co-operative Printing Union has had the highest visibility from its inception to this day because of its very considerable output in all forms under more than a dozen variant imprints.[1] Whether a hardbound book, a legal brief or a commercial billhead, the work represented an excellent professional standard that was competitive in San Francisco.

After the firm's uncertain beginning under Agnes Peterson and its reorganization and promotion by Emily Pitts Stevens, the person completely responsible for eighteen years of success was Mrs. Lizzie G. Richmond. She arrived in San Francisco sometime before the December deadline and was listed in the 1869 Directory as "Manager, Women's Co-operative Printing Union." Although Mrs. Pitts Stevens was still president in 1871 and 1872, Mrs. Richmond had now assumed also the duties of secretary and treasurer, making her the person in charge of the whole operation.[2] Because the WCPU was a public stock company, Mrs. Richmond's move toward ownership was undoubtedly consummated through gradual share purchases so that, by 1877, the firm advertised as "Women's Co-operative Printing Union, Mrs. L. G. Richmond, Proprietor." Two years later, the ownership became "Mrs. L. G. Richmond & Son" but the corporate designation was never abandoned.[3]

Where or how Lizzie Richmond learned the printing business is not known by her descendants nor shown by any existing records.[4] She was born Eliza Gray Brown on 30 October 1837 in Little Compton, Rhode Island, a small community. Pawtucket, the nearest place with printing, had a newspaper, book-and-job-shops, and a "Print Works School, Lydia Payne, Teacher."[5] There is no known connection between any of these establishments and Eliza Brown. One thing is certain: by the time she arrived in San Francisco, she was competent in the business of printing and quickly found employment with the WCPU.

She grew up in Little Compton and was married there to Preston B. Richmond (b. 1832), a native of Savannah, Georgia,

Fig. 24: *The Richmond family home at 518 William St., Oakland.*
Courtesy the Heatherly Family.

Figs. 25 & 26: *Willard P. Richmond, left, and Mary H. Richmond, right.*
Courtesy the Heatherly Family, Oakland.

on 5 April 1855. A son, Willard P. (C 7), was born in Rhode Island on 15 April 1856, followed six years later by Isaac L. (C 6) on 12 January 1862. For a brief time, her husband was a merchant in Benicia, California, but the couple apparently divorced in 1869 or early 1870, and Preston Richmond returned to Rhode Island and remarried on 8 June 1870.[6]

Mrs. Richmond continued her work and lived with her sons at various San Francisco addresses until 1875 when she bought a home at 518 William St., near Telegraph Ave., in Oakland and then commuted to San Francisco daily.[7] Both sons learned printing under the watchful eye of their mother in the plant she managed. Tragically, Isaac died when he was only seventeen and working at the WCPU, so it was left to Willard to assist his mother for many years.

The 1870 Census, which covered the twelve months to 1 June 1870, was one of the most complete ever done and gives many unique details about both individuals and businesses. For example, this was the only time that an eleven-year-old girl

117

would be listed with an occupation, in this notable instance, "apprenticed to printer" (A 156). Information on the WCPU and other printing-offices is detailed and affords a revealing assessment of them. At the WCPU, there were seven females over fifteen years of age along with three men over sixteen. They earned a total of $8,060 in wages during the twelve-month period. Materials on hand included 480 reams of printing paper, worth $2,400; card stock, $1,400; and miscellaneous papers, $360. The gross income was $13,300 which put the WCPU ahead of only a few competitors, but Spaulding and Barto, job printers with the exact same kind of equipment as the WCPU, grossed $18,000 for the same period, while employing only three men and one juvenile. Edward Bosqui, a leading printer of this era, reported gross sales of $40,000 in his printing operation (he also ran a bindery, enumerated separately) while employing fifteen men, four females over fifteen years of age and two juveniles. A. L. Bancroft & Co. and Francis & Valentine, with incomes of $75,000 and $105,000, respectively, were the biggest printing operations in town.

Under the heading of "Power," the Census enumerator noted "hand" for the WCPU plant. The paper- and card-cutters listed were hand-operated devices but the presses were standard Gordon jobbers and thus would have been foot-, i.e., treadle-, operated in a shop lacking other power. These were described in the old manner: $1/2$ medium, $1/4$ medium, and $1/8$ medium. In today's terms, these would be 12" x 18", 10" x 15" and 8" x 12" job presses, although in fact they were slightly larger.[8] A wood-engraving used in many WCPU advertisements clearly shows a treadle-operated press. The WCPU billheads of these years declaimed: "Women Set Type! Women Run Presses!"

Most competing shops in town had similar equipment even when they owned a cylinder press which could handle larger work. All printers were thus vying for the lucrative job work in a growing city, and the WCPU was in a basic position to compete for it. Interestingly, proprietors of the shops met and said "the competition caused by [the new] railroad connection with the East, makes a reduction of prices in California inevitable" and meant a revised wage scale was needed.[9]

Despite the competition and financial vagaries of the time, under Mrs. Richmond's astute management the WCPU gradually succeeded. Writing many years later, Charles A. Murdock, another of the noted printers of her era, recalled that she "conducted it well. She was assisted by a capable son and was courteously treated by her competitors. She was successful in a moderate way and built up a good business . . . ."[10] Times were so difficult that there were more than 150 unemployed printers in San Francisco in late 1869, "and the Typographical Union has taken measures to transport to the East all of its members who desire to go."[11] Despite the economic

Fig. 27: *Mrs. Richmond bought a press like this one to augment her pressroom capabilities. The ad says it "runs easy by hand power."* Courtesy The Bancroft Library.

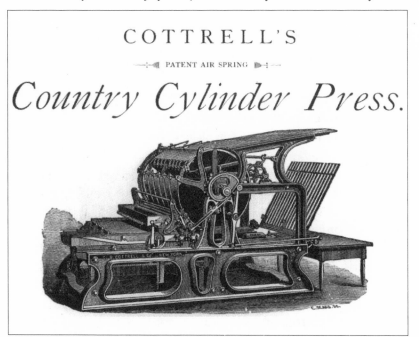

COTTRELL'S

PATENT AIR SPRING

Country Cylinder Press.

slowdown, the Union went on strike the following August for an increase in wages from 60¢ to 75¢ per thousand ems. "The *Call's* publishers refused to accede to the demand . . . and advertised for a new set of hands, which they speedily obtained."[12] This strike finally opened up the lucrative newspaper typesetting jobs to women although, as noted earlier, women always constituted a very small minority of the work force in the larger establishments.

"On the first day of the strike," the *Call* reported, "we visited the manager of the 'Women's Co-operative Printing Union' printing establishment, with the desire of sending 'copy' to that office, and having several columns of *The Call* set daily by the woman compositors there. The lady in charge of the Union was well pleased at the prospect of a large amount of steady and lucrative work for her office, and at once agreed to do the work required. But upon sending to her foreman to arrange details, he proved to be not a 'Woman's Union' man, but a member of the Typographical Union and he flatly refused to set any type, even in their own job office, to be used in *The Call*. We thought that a queer kind of service to the ladies to refuse them an opportunity to earn money when it was offered, and we have not yet discovered how an establishment conducted on such principles is going to render much real service to the unemployed female printers of this city. We have five female compositors employed on *The Call*, whose quiet, lady-like demeanor, and the neat, clean and rapid manner in which they do their work, has earned the commendation of their masculine associates and of the proprietors; and they seem to consider that *The Call* is rendering some practical service to women's rights in offering them an equal opportunity with the stronger sex to earn a support for themselves, a work which they are handsomely accomplishing."[13]

Who the recalcitrant foreman was, and how much longer he remained with the WCPU after this incident, are not presently a matter of record. In response to a *Vallejo Recorder* attack on its policies, the *Call* again pointed out how the Typographical Union had deterred them from hiring women in the past and declared: "The only persons [now] employed . . . came in voluntarily and applied for situations. And lastly, the women employed in *The Call* office — five out of some twenty-five printers — are paid the same price per thousand [ems] as the male compositors." The *Call* concluded: "The truth is, that the *Bulletin*, at one time, had some of its composition done by the women of the [Women's] Co-operative [Printing] Union at their own office, sending 'copy' there to be set up, paying them regular prices for the work. But the Union printers took umbrage at this, and the foreman was actually compelled to stop giving women work, on pain of being driven from his situation."[14]

The 1870 strike lasted eleven days and on 13 August the Union men acknowledged failure and were willing to accept their former rates of pay, 60¢ per thousand ems for an afternoon newspaper and 65¢ for the night work required by a morning newspaper. Most importantly, women had finally gained a permanent foothold in the larger shops in San Francisco as a result of this ill-considered strike which completely negated the Union's long-held power, especially the exclusion of women from work at the better jobs.

Later, the *Bulletin* elaborated on the situation: "Several young women accepted positions on the *Morning Call* a few months since [during the strike] when it was impossible to secure enough male compositors to issue the paper. The female compositors have proved faithful and efficient, and the seven now employed set type with remarkable rapidity and perfection, considering their limited experience.

"The intelligence and capacity of California girls is demonstrated again by the success of the Women's Co-operative Printing Union, No. 424 Montgomery Street. As originally started, the office did not achieve the success anticipated by its projectors, and the first woman [Agnes Peterson] who attempt-

INCORPORATED 1887 UNDER THE LAWS OF THE STATE OF NEVADA

NUMBER
10

SHARES
50

CAPITAL STOCK
$100,000

PAR VALUE
$4 EACH

Rae's Electric System Company

This Certifies that *A. Levy*

is entitled to *Fifty* Shares of the Capital Stock of

Rae's Electric System Company, for State of Nevada alone,

transferable only on the books of the Company by the holder hereof in person or by

Attorney on surrender of this Certificate.

Dayton, Nev. April 14, 1887

C. J. Furman
TREASURER.

Julio Rae
PRESIDENT.

STOCK FULL PAID

WOMEN'S PRINT. S.F.

ed to superintend its affairs retired in disgust. Mrs. Lizzie G. Richmond next assumed control of the establishment, and under her efficient management it is making excellent progress. The office is complete in every respect, and turns out job work which will compare favorably with specimens on exhibition at leading offices of the city. The office gives constant employment to sixteen persons, ten of whom are females. If any expert in the printing business is inclined to think a woman is of little value as a 'type setter,' a visit to this establishment will explode his absurd theory of the question. Most of the compositors are under 20 years of age, we should judge, and have learned what they know of the typographic art in California. There is one young lady — married — occupying a case [stand] in this office whose rapidity and correctness would excite surprise in any office in the land." The article concluded, "Women are peculiarly adapted for the printing business and now that they are showing an interest in the typographic art, it is to be hoped that they will receive the proper encouragement."[15]

One place where they found it was the *San Jose Weekly Mercury*. The editor, James J. Owen, suffragist, Spiritualist, former state legislator and an experienced newspaperman, sold his partnership in the job-printing business of the *Mercury* in November 1870 and took sole control of the newspaper. Fully aware of the WCPU and other female struggles for work as typesetters, he wrote: "Under our new management we propose to try the experiment of introducing girl compositors into our type room. To this end we have commenced with two young women, daughters of the editor [A 198, A 199], and propose to impart to them a thorough knowledge of the printing business. Commencing on Monday last [31 October 1870], and working only seven hours, they learned the boxes and set up of solid brevier [8-pt.], the eldest four 'stickfuls' — over a thou-

Fig. 28, facing page: *One example of work done for State of Nevada customer.* Author's Collection.

sand ems — and the youngest, three. The 'matter' was set up nearly correct. For a first day's work we think that is hard to beat."[16] For several weeks thereafter, fellow editors wrote stories chiding Owen because they believed he had exaggerated the girls' capabilities. Nevertheless, his "experiment" worked and he continued to employ woman typesetters as did his competition. (See A 23, A 110, A 200, A 261.)

Another newspaper noted: "The composing room of the *San Jose Mercury*, since the introduction of *lady compositors*, has been changed from the usual disordered condition of country offices to a perfect model of comfort and neatness. No pipes, no tobacco, no lager beer, and no boisterous language or actions are met with in these rooms."[17] J. J. Owen's correspondent in San Francisco immediately followed up: "I no sooner read [about it] than I resolved on an exploring expedition, wishing to see what was doing in other offices. Accordingly, on Saturday last [17 December 1870], I armed myself with paper and pencil, climbed the fearful heights[18] leading to the rooms of the 'Woman's Co-operative Printing Establishment [sic],' and introduced myself to Mrs. Richmond, manager of the place. She received me pleasantly, and gave me the following information: There are at present ten women steadily employed; four of these are apprentices, and receive no pay for the first three months, after that they earn fifty cents a thousand [ems].[19] Two of them are first class operatives, and receive twenty-five dollars a week, while the remaining four earn fair wages. The chief trouble, heretofore, has been that girls, so soon as they have mastered the business, accept situations in other places, thereby leaving the Incorporation to the mercy of apprentices; and you know what kind of proof fresh apprentices will present you for correction.

"The institution has been incorporated little more than a year now, and is now established on a firm and paying basis. There is never any lack of job work, but often of matter for composition [i.e., straight matter], and this latter is what is especially needed in such an office."[20]

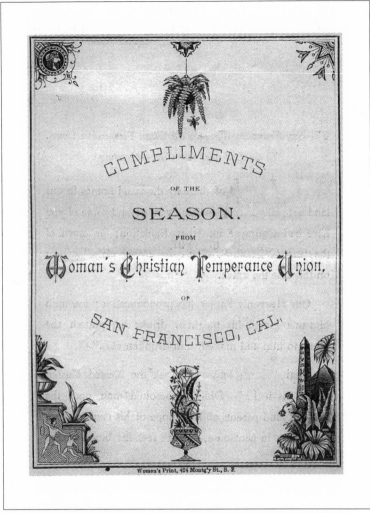

Fig. 29: *The W.C.T.U. was organized after Emily Pitts Stevens left the suffrage movement to become a temperance lecturer.* Author's Collection.

Three points should be noted: first, that women could not make enough money on the thousand-em basis in a shop where jobs consisting of only a few lines (compared to book or newspaper work), were the primary typesetting and where the

125

basic pay was already lower. This is why the newly-trained often sought work elsewhere and why the opening of newspaper typesetting to them was crucial. Second, those cited as earning $25.00 were in fact "compositors" and had completely different duties as opposed to typesetters who performed one task only and were paid for piece work. Third, it was a general practice in San Francisco to pay nothing to apprentice typesetters for the first three months.[21]

Another document affords a personal connection between Lizzie Richmond and her thriving business. On 12 May 1879, she wrote to her son from San Diego: "My darling boy: How are you getting along at the office, I have not given it one thought or care since I left, but I hope it will go all right — any way I am in a fair way not to know much about it, as I have not get any mail yet & it is a week today but it seems a month in one sense . . . .

"Give my love to all hands at the office [and] tell them that I am getting a coat of tan . . . ."[22]

During the 19th Century, a title page did not necessarily reflect the publisher of the work, only the printer, and as a result the WCPU imprint appears in a wide variety of books from its beginning (see Appendix B). An early and most intriguing title was: *One of the Cunning Men of San Francisco; or, Woman's Wrongs: Being Sketches From the Diary of a Neglected Wife* (1869, B 27), with the "I" in "CUNNING" upside-down on the title page. Another provocative example was: *The Polygamist's Victim; The Life and Experiences of the Author During a Six Years' Experience Among the Mormon Saints, Being a Description of the Massacres, Struggles, Dangers, Toils and Vicissitudes of Border Life, With Miraculous Preservations in Many Dangers by the Intervention of Mysterious Powers* (1872, B 67), surely one of the longest and most dramatic Western title pages of all time. This was followed by *Six Years Experience as a Book Agent in California* (1874, B 84) which has many vivid woodcuts but lacks the professional competence of most of the other books from this period.[23] Some titles suggest that the woman's activism of the time may have encouraged their authors to seek publication.

Over the years, the WCPU printed books with a very wide range of content including two cook books (B 35, B 54), a children's book (B 126), a Spiritualist title (B 126) by Cora V. Richmond (pseud., no relation to Mrs. L. G. Richmond), more than any other woman-operated printing-office. In the long run, however, it was the legal briefs, annual reports, constitutions-and-by-laws and the large numbers of billheads, flyers and other utility printing that made the shop profitable. Although an 1877 advertisement announced "Printing in Spanish a Specialty," no WCPU Spanish imprint has surfaced despite intensive search.[24] Mrs. Richmond also sought work from out of town and did a great deal of work for businesses in outlying California towns and as far away as Nevada.[25]

A new feminist publication, *Woman's Pacific Coast Journal*, edited by Mrs. Carrie F. Young, first appeared in May 1870, printed by the WCPU in a neat, three-column layout. It ran to sixteen pages per issue and undoubtedly its format was easily printed *two-up* — two pages at a time — on the $^1/_2$ medium (12" x 18") press. The editor reported: "Visiting the composing room of the Women's Co-operative Printing Union, we approached the case where Susie Moran [A 180] and Sarah Fisher [A 84] were at work. Our salutation interrupted the steady click, click of the types their nimble fingers were placing in the composing sticks.

"'We hope,' said they, 'you will like your paper, for we have set most of the type for this number.'

"The girls are but fourteen years of age, and as they are well behaved, industrious, and good, we take pleasure in acknowledging our obligations to them for the neat appearance of the present number."[26]

The 1870 Census showed Sarah Fisher, a native of New York, as 16, rather than 14, while the 1880 Census gave the age of Massachusetts-born Susie Moran as 23, making her only 13 in 1870. Both were apprentices at the WCPU and Ms. Fisher dropped from sight after the 1873 Directory. However, Susie Moran is notable for her longevity as a typesetter because she remained in the trade from 1869 until 1886, becoming a member of the San Francisco Typographical Union in 1883 when the ban against women members was finally lifted. She worked at *The Call* from 1875 to 1886 and then became an Exempt Member (non-working) of the Union to join her mother and a sister in a millinery business in Oakland. (See A 180 for complete details.)

On another occasion, Carrie Young wrote, "The type for this number [December 1870] of our *Journal* has been set by a woman, who sets ten thousand ems in eight hours — our crooked manuscript copy — and shows a very clean proof."[27] This was first-class typesetting by any standard and the resultant pay of $5.00 for the day's work put this worker ahead of many of her male colleagues in the city.[28]

J. J. Owen thought the *Woman's Pacific Coast Journal's* "typographical appearance is faultless."[29] Despite the added praise of others, by February 1871 the magazine had a new printer, Fish & Co., 511 Sacramento Street. Fish's *Journal* advertisement was headed "Women's Oriental Printing House" and said: "In the employment of hands preference in all cases is given to women. We now give steady work to four, one of whom, if not THE BEST, is ONE OF THE BEST hands we ever saw in a printing office. As we are printing two papers, and have a good line of Book and Job Work, we can give employment to four more women compositors on application."[30] This was another example of the growing opportunities for woman typesetters once they had been trained.

The accelerated growth of San Francisco brought a corresponding increase in printing activity and many supporting businesses, such as type foundries and paper mills, which were successful in filling many needs locally and cutting dependence on the long haul from the East. Thus William A. Faulkner & Son reported casting 30,000 lbs. of type in its first year while the Pioneer Paper Mill in Marin County made 46,000 lbs. of printing paper. In 1870, San Franciosco had thirty-eight printing-offices with 400 printers at work.[31] Faulkner also had the frames of ten Washington handpresses cast by a local iron foundry.[32]

As printing grew in all aspects, trade publications also appeared, usually issued by type foundries. References to Mrs. Richmond and her WCPU appeared passim, including a flattering comment about a Memorandum Calendar, "the neatest

Fig. 30: *Spiritualist editor and publisher.* From The Carrier Dove. Courtesy The Bancroft Library.

and most compact thing of its kind we have ever seen. It contains space for memoranda ... fire alarm, ferry time tables, rates of fare for all parts of the coast ... etc. Under Mrs. Richmond's able management, this office is taking high rank, and we are pleased to chronicle its continued success."[33]

That Mrs. Richmond was aggressive in seeking work is shown by her winning a contract from the San Francisco Board of Supervisors worth $974.50 for "printing Supplements to Ward Registers. . . ."[34] Further insight into her successful business efforts appears in surviving paper-company records which list monthly purchases and outstanding balances yearly from January 1876 to May 1888. For example, 1876 acquisitions totaled $6,544.74, an average of $545.40 per month. The same ledger shows these figures are comparable to the amounts spent by Edward Bosqui & Co., one of the city's leading printers.[35] Mrs. Richmond had an outstanding balance of $2,275.84 at the end of 1876; paper companies have tended to "carry" their customers and thus were, in a sense, the financiers of printers.

By a crude estimate, based on the 1870 Census figure of $5.00 per ream cited above without differentiating sizes, Mrs. Richmond used some 1,309 reams or 654,000 sheets of paper, a very respectable amount from a single source for the period. Over twelve years, her monthly purchases varied (there may have been other suppliers also) from a low of $57.72 to a high of $809.18.[36] Principally, the Ledger shows a consistent business, which reflects credit on Mrs. Richmond's astute management. A trade reference work also listed her credit as "good."[37]

A developing printing business frequently requires new equipment both for expansion and to handle better work, already at hand. Lizzie Richmond was no exception. There was a constant need for new type and the house organ published by Painter & Co., San Francisco typefounders, noted, "Women's Co-operative, new job letter," and on another occasion, a "neat assortment of Display Type," had been acquired.[38] In general, the WCPU's types were much the same as those in daily use in other San Francisco printing-offices including many ornamented faces of varying quality. However, Mrs. Richmond also bought Caslon and used it now and again effectively.

Later, a very important acquisition was reported: "The Women's Co-operative Printing Office, under the management of Mrs. L. G. Richmond & Son, one of the leading offices of the city in the production of first-class commercial and artistic work, has just added a new 31" x 46" Cottrell cylinder to its pressroom."[39] The dimensions are for the bed that holds the type and a sheet as large as 28" x 42" (a standard size at the time) could be accommodated effectively. Consequently, a regular 6" x 9" octavo book could be printed *16-up* — sixteen pages at a time — a cost-effective procedure. The new press represented a basic investment of $1,100.

A major project for this press came to the WCPU when Mrs. Julia Stevens Fish Schlesinger (A 232), editor and publisher of *The Carrier Dove*, a Spiritualist weekly which had been primarily for children, decided to expand her Oakland magazine and to publish in San Francisco, "feeling that the change would be beneficial." Mrs. Schlesinger added: "*The Dove* comes out in a new dress, and in our opinion, a finer one than heretofore. It will be printed by the Woman's Co-operative Printing Company [sic], under the management of Mrs. Richmond and Son, who are so well known on this Coast as to require no praise from us. The work speaks for itself."[40] *The Carrier Dove* now appeared in a three-column, 8$\frac{1}{2}$" x 11" format which was easily printed *8-up* — eight pages at a time — on the Cottrell cylinder press. It was not only well printed but Mrs. Schlesinger was also a first-class editor. Despite a basic preoccupation with Spiritualism, no 19th Century publication before or since Emily Pitts Stevens's *Pioneer* contained more feminist news or such a heavy emphasis on suffrage. *The Carrier Dove* established itself nationally in a very short time as well.

A romance evolved from this move to the WCPU where one of the typesetters was Miss Nellie Gorman (A 106, A 107). She decided to join Mrs. Schlesinger who announced "that we are now established in our new office at 841 Market Street, where hereafter the *Dove* will be printed."[41] Here she met a fellow-employee, Milo Fish, Mrs. Schlesinger's son by a previous marriage, and they married. Nellie Gorman continued to work for many years as a typesetter despite — and possibly because of — her growing family. Interestingly, by August 1888 Mrs.

Fig. 31: *Nellie F. Gorman, typesetter.* Courtesy Michael Engh, S. J.

Schlesinger had also added "a large, new cylinder press" to existing equipment and the following week could announce that *The Dove* "has been printed on our own machine."[42]

With Lizzie Richmond's business a success and her ownership complete, she remarried. Her new husband was Norris A. Judd, a native of Pennsylvania, then 23. At the time, he was in the classic American pattern of making his way upward in an established business, the D. Hicks & Co. bookbindery, in which he became a partner in 1879.[43] For the marriage certificate, Lizzie Richmond declared her age as 35 when she was actually 40. They were married on 4 August 1877 in Oakland by the well-known San Francisco Unitarian minister, Dr. Horatio Stebbins, and the couple lived thereafter in the William St. home for many years.[44]

Meanwhile, son Willard also had taken a bride, Mary H. Shearer, a native Californian, and 1880 Census data show both families living at the William St. address and Norris Judd only two years older than his step-son. The household also included Lizzie Richmond's year-old grandson, Lester, and Ding, 25, a native of China and live-in servant. Willard and Mary later had a daughter, Edna, but there was no offspring from Lizzie Richmond's second marriage.

In later years, such as 1886, Mrs. Judd listed herself three ways in the San Francisco Directory: "Mrs. L. G. Judd; Mrs. Lizzie G. Richmond-Judd; and Mrs. L. G. Richmond & Son (Mrs. L. G. Judd & Willard P. Richmond, proprietors, Women's Co-operative Printing Union)." These variants assured maximum publicity for her firm and helped to reduce any confusion that might have resulted from her change of name. Personal and business matters remained as before, however, until sometime in 1887 when the Women's Co-operative Printing Union moved to 23 First St., to be merged with Hicks-Judd's already long-established printing department and the bookbindery. As of the following March, the labor force of Mrs. L. G. Richmond & Son consisted of six males and one female who were

Fig. 32, left, *shows merger of the two firms by rubber stamp.* Fig. 33, right, *shows the merger was completed by the formation of a new corporation. Note that the woman's firm retains its identity, with a slightly altered name, in panel on right.* Author's Collection.

Union members; there was one non-Union male, and six girls and one boy were apprentices.[45]

A new corporation, The Hicks–Judd Company, formalized the merger on 21 May 1888 "for publishing, printing and bookbinding," with Norris A. Judd (370 shares, $37,000); Mrs. Lizzie G. Judd (25 shares, $2,500); Daniel Hicks (5 shares, $500); Charles K. Rosenberg (50 shares, $5,000); and Willard P. Richmond (50 shares, $5,000) as principals.[46] Only the 1891 Oakland Directory listed Lizzie Judd as vice-president of The Hicks-Judd Co., but she probably served in that position from the incorporation, just as her husband was president from the outset. The new arrangement was very successful and all participants evidently shared generous dividends. Thus Eliza Judd retired comfortably after eighteen busy and productive years at the helm of the WCPU. The merger also assured continuing employment to those men and women who still wished to work.

The WCPU, a well-established local business for so many years, remained an identifiable entity under the new arrangement. Billheads and Directory advertising also noted, "Successors to the Women's Co-operative Printing Union."[47] The WCPU retained this identity until 3 September 1901 when a fire broke out on the top floor of the building at 23 First St. which "did considerable damage to the plant and building." The shop and stock were valued at $150,000 and the company had $70,000 in insurance so were soon able to resume operations. The WCPU remained in the ashes, never to rise again.[48] The Hicks-Judd Co. was an increasingly successful business until Nelson Judd sold out to the Sunset Publishing Co. in 1920. He had remarried in 1909 and died 11 January 1937.[49] None of his obituaries mentioned his first marriage.

Eliza G. Richmond Judd died at age 61, "after months of suffering with cancer," on 8 October 1898 while again visiting in San Diego.[50] Her body was returned to Oakland for internment in the family plot in Mountain View Cemetery on 15 October. Services were held at the First Unitarian Church which "was filled with a large number of friends and relatives . . . . The floral pieces sent by employees of The Hicks–Judd Company were beautiful. . . . The employees of the job and press department contributed a large raised floral wreath with a sheaf of wheat."[51]

A story in the *Oakland Enquirer*, where son Willard was working as a compositor at the time, contained several misstatements of fact and also had some gossip. "Her marriage with Judd, it is said, did not prove a happy alliance, and they were living apart at the time of her death." Further, "Mrs. Judd had lately become a believer in Spiritualism," and, finally, "She was in receipt of a regular income from the printing company."[52]

When his mother died, Willard had not been in the employ of The Hicks Judd Co. for some time. Apparently, problems with alcohol led to his leaving, so that by 1893 he had a printing business with his wife (A 220), W. P. Richmond & Co., at 514 Stockton St., San Francisco. The firm is unlisted after 1894 in any city Directory and by 1896 the Oakland Directory shows him as "Pacific Coast Agent, The Physicians' and Surgeons' Soap." Printing evidently was no longer a part of his life. He died prematurely 1 March 1899, at age 41, of cirrhosis of the liver.[53]

The large assortment of printed materials created by the WCPU, now in many California libraries and private collections, provides the evidence that Eliza Richmond Judd was a very competent master-printer. She also contributed to the success of women in printing in San Francisco by consistently training and employing them, although she left no written record of her views on women in printing or the feminist problems of her day.[54]

Chapter 8 Two Other Unusual San Francisco Women

Amanda M. Slocum, Spiritualist, suffragist, editor and finally a master-printer of achievement, first came to the attention of San Franciscans when she spoke as a delegate from San Jose at the first Pacific Coast Woman Suffrage Convention, held in San Francisco, 17–18 May 1871. According to the *Call*, she "read a lengthy if not particularly brilliant or original essay on the subject of Woman Suffrage" to a sparsely attended second session of the convention.[1] She gained prominence in both publishing and printing after moving to San Francisco, where she and her husband, William N. Slocum, began to publish *Common Sense,* "A Journal of Live Ideas," 16 May 1874.

She was a native of Iowa, and the first record of her in California is the 1860 Census for San Jose where she was listed under her married name of Amanda Taylor (A 268) with an Iowa-born daughter, Etta M. Taylor (later, Eva T.; see A 241), five months old. The San Jose household in which she lived was headed by a woman, "H. A. Anderson," 48, and included siblings 25, 16 and 12 years of age, all born in Iowa.[2] At age 20, Amanda fitted the spacing pattern of the Anderson children and she very likely had been an Anderson before her first marriage. She was also a widow at this young age.[3]

Where and how Amanda Taylor and William Slocum met and courted is unknown. The few facts about her indicate very little experience in any direction at this time, but it is possible that she may have been working in the printing trade in San Jose and met her future husband when he moved there. William was a multi-talented individual with much experience in both journalism and printing. According to information he furnished for his family's genealogy, he had first come to San Francisco in the late summer of 1856 to cover the Vigilance Committee for the Associated Press, but the Committee had disbanded before his arrival.

In 1857, he bought an interest in a placer gold mine. This could hardly have been a success because the Whitton & Towne (Towne & Bacon) records at Stanford University show him earning $10.80 for the week ending 8 May 1858, most likely for typesetting, at which he was adept.[4] The next year, Slocum was living in Santa Cruz, and on 29 August 1859, bought out the weekly *Santa Cruz News* as an organ of the fledgling Republican Party. However, it lasted less than a year, dying on 21 July 1860, due to Slocum's "heresies," according to a rival newspaper. San Jose was more to his liking, and Slocum fit well into the Republican and Spiritualist community there. In August, he used the equipment from the *News* to start the *San Jose Telegraph*, shortly renamed the *Mercury*. However, he continually sought political office, and in May 1861, moved to the San Francisco Custom House, leaving the *Mercury* to J. J. Owen. In November, Slocum published *The War and How to End It*, the first California pamphlet to advocate the emancipation of southern slaves.

The 1862 San Francisco Directory next listed him as a

Fig. 34, facing page: *A neatly done legal brief printed by Amanda Slocum's newly incorporated firm*. Author's Collection.

No. _4 8 0 0_

IN THE

SUPREME COURT

OF THE

STATE OF CALIFORNIA.

ELLEN REIMER, Respondent, *vs.* EDWARD L. REIMER, Appellant.

POINTS, Etc., ON PART OF RESPONDENT.

LOWELL J. HARDY,
Attorney for Appellant.

M. G. COBB,
Attorney for Respondent.

Filed .... 1875.

.... *Clerk.*

Received a Copy of the within Points, etc., this .... *day of*
.... 1875.

....
Att'y for Appellant.

SAN FRANCISCO:
Woman's Printing Association, 605 Montgomery Street.
1875.

printer, a trade he could always fall back on for survival. This was also the year he and Amanda were married "in San Francisco." He evidently adopted her daughter and they had two more children, Clara (A 240), born 28 February 1863, and Frederick, born February 1864 and dying prematurely at age four.[5]

(By coincidence, another Amanda Taylor (A 267) was working in printing in San Francisco at this time and appeared in the 1875 Directory as a compositor at the *Occident*. She has a small claim to fame as "the first young lady amateur [journalist] of the Coast."[6] Her publication, *The Olive Branch*, reflected her expertise both in writing and in the printing of its four-page format. She noted that "eight papers are edited and published by boys" but there was none by "even one of California's fair daughters." She said: "[N]ow we intend to try and supply the deficiency by bringing before the public a paper which *a girl is to edit and publish*, and which is to be contributed to, as far as possible, by girls."[7] There is no record of the number of issues Amanda Taylor published.)

Tracing Amanda Slocum's life after the marriage is especially difficult because her husband was changing jobs so frequently. In January 1862, he was a candidate for Clerk of the Assembly; he sought the Collector of Customs job at Monterey; he was Assistant Clerk of the Assembly in 1862 and 1863; he became Inspector of Customs at Port of Santa Cruz; and, in 1865, Collector of Internal Revenue for the Second District (Santa Cruz South). By 1868, though, Slocum was manufacturing bed springs in San Francisco. With such a varied career, no wonder Slocum had so many "heresies." One "heresy" was rather unusual: "The question 'Why do not printers succeed as well as brewers?' was thus answered: 'Because printers work for the head, and brewers for the stomach, and where twenty men have stomachs, but one has brains.' "[8]

William N. Slocum was a political radical by any standard. In 1876, he prophesized that "The next ten years will bring about a revolution in this country. Society as constituted is a sham; our social system is the upholder of injustice, the bulwark of great wrongs. It is to be reorganized, and present appearances indicate the road it must travel will be bloody."[9] Two years later he also published a pamphlet bearing the title of *Revolution*, in which he declared, "I believe in political communism — government by districts — and, to a certain extent, in social communism — the common holding of property by communities."[10]

Curiously, this 1878 tract also had an "Employment wanted" section in which he stated his wishes and qualifications: "I desire to make an arrangement with a good business man (a printer) for the publication of a reform journal in this city, or a country newspaper anywhere on the Pacific coast. I am a printer by trade, and familiar with all branches of journalism; so, although I have no capital, and little of the money-making faculty, I can be of essential service in company with a good business manager.

"If such an opportunity cannot be found, I would accept a position as editor or reporter on any decent newspaper. Heretofore I have failed to retain such positions on account of ill health; but my health is now good. I abhor tobacco and whisky, and will not have a partner who uses either."[11] Ironically, Amanda Slocum met all the qualifications he sought but *Common Sense* was long since dead, in no small measure because of his "heresies," and they had not been together for many years at the time the booklet was published.

The young Amanda Taylor apparently was swept off her feet by this talented and persistent man who was twelve years her senior. *Common Sense* came into being when William Slocum "resigned a position on the *Evening Bulletin* worth $150 per month, sacrificed the savings of years, and gave much arduous labor in the attempt to establish a Spiritual [sic] journal in San Francisco; and Mrs. Amanda M, Slocum, with noble self-sacrifice, gave the entire proceeds of the sale of a herd of valuable cattle, which, together with cash contributed by her,

Interest will be charged at the rate of 1½ per cent, per month on all bills, if not paid at the expiration of sixty days from date of invoice.

Payable in U. S. Gold Coin.
Silver, Currency and Gold Dust taken at Market Value only.

San Francisco, _Decr. 20_ 1878

M_____ C. S. Healy

Bought of Glasgow Iron and Metal Importing Co.

Nos. 22 and 24 FREMONT ST., near Market Street. (P. O. Box 711.)

Boiler Flues, Sheet and Plate Iron, Rivets, Carriage Bolts, Gas and Water Pipes, Gas and Water Fittings, Spring and Cast Steel, Pig Iron, Anvils, Rasps and Files, Horse-Shoe Iron and Nails, Best Welding E Cumberland Coal.

W. McCRINDLE, Manager.

Notes drawing Interest must be given for all accounts, if not paid when due.

ANANDA M. SLOCUM, Printer, 534 Commercial St.

1 Sheet Iron 8 × 30 #12 82 4½

Pard Glasgow off the date #840

amounted to over four thousand dollars, and all she, as well as I, got out of it was a most bitter experience . . . ."[12] The statement suggests that the whole enterprise depended on the huge investment made by Amanda Slocum, in what was probably her husband's idea in the first place, based on his previous and later experiences. Where or how she acquired such large assets are not a matter of record but by inheritance might be a fair assumption.

Their journal began its weekly run 16 May 1874 with husband and wife listed as editors. Eventually, Mrs. Slocum became assistant editor and business manager. In this first issue, they said "the beautiful white paper" on which *Common Sense* was printed came "from the Pioneer Paper Mill [Marin County] of S. P. Taylor & Co." They also noted, "It lacks the finish of the sized and calendered paper of Eastern make, but in other respects it is superior; and we believe in patronizing home institutions."[13]

From the outset, suffrage, temperance and other women's issues were nearly as prominent as Spiritualism, the "liberal" base for the publication. A good example is a letter from Carrie F. Young from her lecture tour of the northern part of the state. "The temperance reform is carried on where I have been by bigoted church people, who freeze me out when they learn I am not one of them. . . . I am surprised that Emily Pitts Stevens is permitted to work with them. Or has she joined the church? [Yes.]" There was a wide variety of reports, ranging from the jailing again of one Jennie Bonnet, "the girl who wears men's clothes while catching frogs," for wearing that apparel to frequent reports on woman suffrage affairs and meetings. Victoria Woodhull's lecture in San Francisco was duly reported but quite differently from the other newspapers.

William Slocum, "with wife and two daughters, went to

Fig. 34, facing page: *Example of billhead produced by Amanda Slocum under her own name.* Author's Collection.

hear the woman who is so nicely abused. . . . When Mrs., Woodhull declared in the most emphatic language against promiscuity, not a reporter made a scratch of a pen. . . ."[14]

The Slocums started to organize a stock company, apparently not recorded with the Secretary of State, in 1874. The initial slate of officers did not include William but Amanda Slocum was vice-president with Alexander Kendrick, a clothing salesman, as president.[15] A month later, some 500 shares had been sold, "We desire as soon as practicable to establish a Job Printing Office and a Liberal [i.e., Spiritualist] Book Store in connection with the paper." [16] As with other independent publications, the Slocums set their own type so the addition of the facilities for job work could enhance their possibilities for added income.

However, it was not until March 1875 that William Slocum could say: "The editor of *Common Sense* is a practical printer, and it has been necessary for him to devote more than three fourths of his time to the printing department of the paper. But after this week, he will be relieved in a great measure from this work, and will be able to give more time to editorial duties." (This explanation suggests that Amanda Slocum had been doing the bulk of the editorial work to this point.) The change was possible because the Common Sense Publishing Co. had "purchased an extensive Steam Printing Establishment, and is now ready to do Book and Job Printing of all kinds . . . . The office purchased is that of the Woman's Publishing Company [Emily Pitts Stevens], formerly located at 511 Sacramento street, now at 605 Montgomery, in the building formerly occupied by the *Evening Post*, where it is carried on under the old name, which is retained because it indicates the fact that the work in the office is performed mainly by women. The establishment is under charge of Mrs. A. M. Slocum, the Business Manager of *Common Sense* . . . ."

The article also noted that the foreman "is one of the best Job Printers in the State, and his assistants are the most capable of either sex that can be obtained." The editor admitted: "We incur, as is usual in business operations, some indebtedness." To keep their good credit he offered stock of five dollars par value for one dollar a share. "Who will volunteer to help us? *Now* is the time. Aid deferred is no aid at all."[17] A new corporation, duly registered, was formed on 26 June 1875 with the name of the Woman's Printing Association to differentiate it from the Woman's Publishing Co., now a part of the Slocums' holdings.[18]

What Amanda Slocum thought about women in printing was more realistic than idealistic. She said: " I know of no good reason why a woman cannot make just as good a job printer [doing layouts] as any man; but so far I have not found them equal to the men even in plain book work [straight matter]. In fact, the most serious difficulty I have yet encountered . . . is that growing out of the incompetency of female operatives. It appears to be the policy — a very bad one I think — in some offices, not conducted by women, to employ girls only so long as they are willing to work at little or no wages, and allow them to go as soon as they ascertain their own value. Girls are also employed who have no natural and little fitness for the trade — some of them destitute even of the rudiments of an ordinary school education. Such a system of apprenticeship is injurious to the girls, and detrimental to the business. It results in crowding the printing offices with applicants for work who are not capable of doing credit to themselves nor to the office that employs them. My plan is to pay women the same wages that men receive, and to employ none permanently who are incompetent. It is an undeniable fact that female type-setters are not generally as efficient as males. This, in my opinion, is not owing to their organization, but to education; it is the effect naturally resulting from the idea of dependence which is instilled into woman by the very atmosphere in which she moves from her childhood. She is taught to look forward to marriage as the end to be attained.

. . . It teaches her to look forward to means of support other than her own exertions, and. . .it impairs her self-reliance and prevents a high degree of skill in her profession. . . . [B]ut in that good time coming, when woman will be on an equality with man politically, then she will be his equal in the field of work as well as in the home, and will have higher incentives to effort than now, and greater encouragement to perfect herself in any chosen art."[19]

Common Sense came to an end with the third number of the second volume, 5 June 1875. William Slocum declared "we do not have the capital to carry on the business" and "we do not feel like forcing upon the public that which is not wanted even by those calling themselves 'Liberals.' " He summarized the problem well when he wrote: "Some 'workingmen' are opposed to Spiritualism; some Spiritualists to Atheism, and some Atheists to Social Freedom, while other reformers pooh-pooh at Astrology, Re-incarnation, Woman Suffrage, or whatever happens to rub against their prejudices. In short, the so-called reformers are discordant classes, too intolerant of each other to agree upon the support of the same journal . . . ."[20]

He also revealed: "The loss on the publication of the paper has never been less than $200 per month. The debts owing by us now amount to about $500." He asked delinquent subscribers to pay up and then said that "Our Book and Job Printing business, with the exception of the *Common Sense* department, has, from the first, been in a flourishing condition." He noted that it had paid "nearly one thousand dollars of the debts of *Common Sense*," and concluded: "The financial failure of the paper is sufficient evidence that it was not wanted; the financial success of the Woman's Publishing Company, under the superintendence of Mrs. Slocum, is equally good evidence that the efforts of women to support themselves by honorable employment are appreciated, and will be sustained by the public."[21]

Both Slocum daughters were taught to set type. Eva, the

elder, was old enough to have worked on *Common Sense* before its demise. Mrs. Slocum started to publish a temperance magazine, *Roll Call*, when the other daughter was fifteen and Clara (A 75, A 240) was credited as "mainly" responsible for editing the magazine as well as with setting all the type.[22] Prior to this, Clara had appeared in print under her own by-line in *Sunshine*, "a magazine for young readers," published in Santa Clara. "The Enchanted Child, A Fairy Story" appeared in the April 1875 issue (p. 114:2) "by Carrie Slocum (age 12 years), San Francisco, Cal." Amanda Slocum was evidently very pleased and ran a one-inch advertisement the following five months for the Woman's Publishing Co., listing herself as superintendent. Interestingly, the April issue also carried a poem, "Women's Rights" (p. 14:1), a topic not usual in juvenile magazines, then or now.

According to the San Francisco Directory, the Slocums were still a family in 1876. Mrs. Slocum was listed as "book and job printing," and he as a "journalist with the *Evening Post*." They resided at 615 Third St. By the time of the next Directory, they were living separately and Amanda Slocum and daughter Eva had new jobs with Taylor & Nevin, a printing partnership from 1877 to 1880. Mrs. Slocum was working as a "solicitor," that is, a printing salesperson, while Eva set type. In view of the success of the Woman's Publishing Co., noted so well by William Slocum, why would mother and daughter have sought work elsewhere? A probable explanation is that in the breakup of the marriage the printing-office was in legal contention. From 1877, Amanda Slocum printed under her own name and never reverted to the imprints of the two corporations with which she had been involved. As these existed — at least on paper — until after the turn of the century, it is very likely the assets had been sold to others in the marriage settlement and that Mrs. Slocum started anew by herself. She was no longer at 605 Montgomery St., where she and her husband originally had moved the equipment of the Woman's Publish-

CUTTINGS:

SELECTED FROM THE WRITINGS OF

MRS. P. ANNETTA PECKHAM,

AUTHOR OF "WELDED LINKS."

PRICE, $1.50.

SAN FRANCISCO :
AMANDA M. SLOCUM, BOOK AND JOB PRINTER,
612 Clay Street.
1877.

ing Co. some years before. Working for Taylor & Nevin gave her daughter and herself a means of livelihood while these other matters were being settled.[23]

Once Amanda Slocum was on her own, she again succeeded in the highly competitive San Francisco of her time. Her few books, both from the Woman's Printing Association days and later, all showed a professional competence and style that compared favorably with the work of her rivals, male and female. One of her more interesting, issued under her imprint in 1877, is *Cuttings*, by Mrs. P. Annetta Peckham (B 107). It has a guarded-in frontispiece of a small cabinet-photograph of the author by Bradley & Rulofson, complete with protective tissue. This was a second edition, the first having been done by Bacon & Co. the same year, and contains two poems and press notices that were new. The body of the book, a mixture of short essays and poems, is well set in 12-point Old Style #2, making it most legible.

An 1881 volume, *Astrea*, by Mrs. E. P. Thorndike (B 159), is another example of a Spiritualist work, this one mostly in the form of poetry. A notable feature is an index, an uncommon occurrence in a text of this kind at this time. The tailpiece, made up of ornamental pieces and type, occurs before the index and says, "Rest artist, thy work is done." At least three variant bindings of this book are known.

Amanda Slocum sustained her strong interest in woman affairs and served as a member of the board of directors of the California State Suffrage Association, Inc., in 1881. She continued her success as a master-printer and last appeared as a "printer" in the 1884 Directory. Later, she became Mrs. Amanda Slocum Reed (possibly Reid) through remarriage and disappeared from the record completely.

Fig. 36, facing page: *Title page of a book produced by Amanda Slocum*. Courtesy The Bancroft Library.

II

Marietta Lois Stow (A 262) was another of the politically oriented women who came to San Francisco and created or worked in organizations that supported women's causes forcefully and also became involved with women in printing. She is best known today, however, for her politics: she was the first woman to run for governor of California (as an independent, 1882); she helped form the Equal Rights Party in 1884 and became its vice-presidential candidate. Belva Lockwood, an eastern attorney of note and the first woman admitted to practice before the Supreme Court, was the presidential nominee. They were the first women to run on a national ticket for these offices, and Marietta Stow also was the first female to chair a national political convention.[24] One standard reference summed up by saying: "Although [the] campaign elicited considerable ridicule and was opposed by Susan B. Anthony and other suffrage leaders, it yet maintained its dignity and generated much public interest. [The candidates] received 4,149 votes in six states and claimed to have been defrauded of more. [Belva Lockwood] ran again in 1888, but the novelty had worn off and results were less impressive."[25]

Mrs. Stow was born Marietta L. Beers in Webster, New York, in 1835 and emigrated with her family to Columbia, Ohio, when she was two-and-a-half years old. She said she only "learned the alphabet after her ninth birthday" because of her mother's illness and the "great distance from school." However, by age fifteen she was a schoolmarm for sixty students. "At nineteen she was married to E. F. Bell . . . a merchant of Cleveland, Ohio, and at the age of twenty-three was a widow and childless." Her only child had died in a scarlet fever epidemic, and after his burial Marietta Bell "fled" to New York "where she soon became identified with humanitarian work." One of her interests was an association for the protection of shop girls, and she began speaking on behalf of her cause in 1859 and became a much-praised lecturer.

Fig. 37: *From* Probate
Confiscation. *Marietta
Beers-Stow was the name
she preferred.* Courtesy
The Bancroft Library.

The last year of the Civil War, "she traveled over 25,000 miles and had subscribed . . . over $50,000 to found a 'National Home and School' combined for the destitute orphan daughters of soldiers. Nearly $8,000 in gold of this money was collected in San Francisco and other California towns, and placed in the hands of the treasurer, J. W. Stow (who was first her treasurer and then her treasure). This money is builded [sic] in a monument to woman's indefatigable energy in the San Francisco Ladies Protection and Relief Society Home." [26]

She married Joseph W. Stow, "a prominent hardware merchant of San Francisco," in 1866. This is the same Stow who later became one of the incorporators of the Women's

151

Co-operative Printing Union and a strong supporter of the woman suffrage movement until he died of tuberculosis in 1874 while Mrs. Stow was abroad. He was a man of means and she was penniless, but because of the laws of the state and the lack of a will, she battled long and hard without success to gain access to some of his assets. This fight resulted in two books, *Probate Confiscation* (1876; 4th ed., 1879) and *Probate Chaff* (1879), both of which sold widely. Mrs. Stow also lectured frequently on the subject of probate and undoubtedly hastened the adoption of community property laws in California .

In 1880, Marietta Stow ran for Director of Schools in San Francisco and organized a mass political rally of women at her home, the first of its kind. She was the second president of the California Woman's Suffrage Association and an organizer and later head of the California Woman's Social Science Association which finally brought her to printing. The Social Science Sisterhood (SSS) within the latter organization enabled her to carry on many of her causes, including "political reform, dress reform and food reform."[27] For example, "Mrs. Stow designed her own clothing, which she called the 'Triple S Costume.' It consisted of a man's pair of trousers over which she wore a kilt skirt."[28] Spiritualism was the one major attraction for so many women of her time that she never embraced. She was, however, an ardent suffragist.

Central to her advocacy was her newspaper, *Woman's Herald of Industry and Social Science Coöperator.* ("There is nothing which the human mind can conceive which it may not execute.") The paper's first issue, September 1881, noted that three departments were to be made special: "Industrial Education; Co-operative Production and Distribution; Ownership of Person, Children, Property." In view of all the so-called "cooperation" in San Francisco prior to this time, it is noteworthy that Mrs. Stow regularly devoted one whole page of her multicolumned newspaper to printing the Rochdale Co-operative Rules, starting in January 1882 (p. 5). Many to this day regard

Fig. 38: *Logo of Marietta Stow's Social Science Sisterhood.*
From Woman's Herald of Industry. Courtesy California State Library.

the Rochdale program as the basis for a "true" co-operative enterprise.

The initial paper also noted that Mrs. Stow was president of the Woman's Social Science Association, 304 Stockton St. (founded 7 August 1880) and that the SSS "is an associate home of *women,* and the first in the United States. This Home, *this dear home,* has a membership of 125 souls." She gave challenging reasons for publishing her paper, for example: "To prove that the triple 'curse' (labor, child bearing and subjection) under which women have been held for all time is false. To preach the new gospel of fewer and better children. To lay bare the mischief resulting from a merely masculine form of government in church and state. To preach the new gospel of Co-operation until the earth is rid of willful and enforced pauperism, until six or eight hours a day will be the maximum of labor." Finally, there came the first of many mentions of printing: "The type from which *Woman's Herald* is printed was furnished by Painter & Co., Clay street, the Pioneer [not true] Type Founders of [this] Coast, who supply everything . . . promptly and in first-class order. The printer of the *Herald*, is H. G. Parsons, 518 Clay street. The appearance of this number speaks for itself as to workmanship of the type founder and the printer."[29] Alas, the good will with her printer was short-lived and the January 1882 number (after missing two issues) told it all: "We read much and often about the wicked plumber, but very little and seldom about the wicked printer, but there is much to be said about both. This pen will scratch the printer. The agreement was (I always make an agreement) that I should have the paper ten days after 'copy' — but instead it came limping in twenty-eight days after 'copy,' accompanied by the delightful promise: 'On time for the next issue.' I washed down my wrath with this drop of consolation, again finished 'copy,' but I wore out all my old shoes and threadbare [sic] patience in looking after that 'copy,' or rather its proof, which looked as though every 'devil' in the establishment had a finger in it. Such somersaulting of letters and pranks of periods I had never dreamed of.

"It took just twenty-one days to make the October number. Then I said, 'Life is too short, time too precious, and good nature too scarce, to put up with this state of things any longer,' and to cap the exasperating climax I found twelve of my chicken's pin-feathers warming another nest — same date, type, everything — and when I 'whyed?' — I always why — was told that it was customary. 'Indeed,' I replied, 'that is a custom I will not put up with.'"

The sequel was: "We have brought the young child home. This number of the *Herald* is manufactured, all but the presswork, under our own motherly eyes. There are two of us, Mrs. [Laura W.] Briggs (A 29) and myself. I have personally set up (first type-setting) all the editorials and the entire article, 'The Second President of the C.W.S.A.' [about herself] and was more pleased with the unique occupation than I would have been in working a whole regiment of blue lines, in Berlin wool, on black-and-tan fustian.

"Typesetting should be largely in the hands of women. It is a light, pleasant occupation and pays well. Women type-setters in this city get all the way from 25 to 30, 35, 40, 45, 50 cts. a thousand ems, and in eight hours' time an ordinary hand will distribute and set six thousand. There is a lady in this city who has set fifteen hundred ems of distributed type in an hour (there is now and then a man who can set two thousand) and another who earns thirty dollars a week type-setting."[30] The wide scale of prices she cited meant that except for the shops of Lisle Lester, Emily Pitts Stevens and — later — the newspapers, most women still were not paid the same as men and consequently were a constant source of cheap labor.

Typesetting had impressed her greatly, and Marietta Stow wrote a most startling comment in the next issue of her paper: "I have never in my life experienced more pleasure — not excepting love-making — than I did in setting the two pages of the January number of the *Herald*, and I never had but one hour's instruction in type-setting. It is claimed by men that women cannot make good typos and successful printers. I do not believe it. Can women learn the art of printing? Come to 304 Stockton St. and see."[31] In this issue she had also shifted her allegiance to Marder, Luse & Co.'s foundry for the "beautiful type (all new) from which the *Herald* is printed. . . Parties about to invest in printing materials will find Mr. [Nelson C.] Hawks, the resident partner, a courteous and obliging gentleman."[32]

This meant, of course, that she now had her own com-posing room and also had started a school for typesetters in addition to the other, ever-changing SSS teaching programs. "It pays better," she said, "to set type than to stand behind a counter and sell dry goods for three dollars a week and it is just as respectable. It is more tiresome to handle ribbons than types and types are the best educators."[33] One of the noticeable consequences of setting type in the SSS workrooms was the many — too many — typographical errors, including a large number of *wrong fonts* — non-matching characters. All are directly attributable to poor proofreading and the employment of so many female novices in typesetting. This newspaper definitely did not meet the standards in San Francisco.

Mrs. Stow's August 1882 advertisement for the "SSS College of Industry for Women and Girls," announced, "Leading Branches taught, Type-setting, Physical Culture, Dress Reform," a highly unusual combination to be included in "industry." The programs had diminished since their inception so that the only other not mentioned in the advertisement, elocution, also seemed to be far from any industrial connotation. But she had been smitten by typesetting, "of special importance, for it is eminently woman's avocation [sic!] and pays far better than slop-work or plain sewing."[34] Other matters were also on her mind in 1883. A front-page editorial declared that castration was "the one radical cure for the crime of rape, incest, and the perpetration of pauperism," a thought heretic to this day. Then she skipped to a more personal matter: "Mrs. J. W. Stow is going to be known by her *Christian* name, Marietta L. Beers-Stow hereafter. . . . She thinks the hard Mr. and the hissing Mrs. and Miss abominable. The simple name, without the ugly handle, is beautiful." (She probably would have vigorously disapproved of Ms., today's "mizzing" sound.)

Then she was back to printing once more. "That women have done well in printing offices, all are agreed. They are far more reliable than the average printer of the other sex, and have had a decidedly beneficial effect in the composing-room, where the deportment of the men has been vastly improved, and the language greatly purified. . . .The suffragists blazed the way. . . ." She also noted that the *Call* and the *Bulletin* "were well pleased with their efficient complement of lady typos" and that the "*Woman's Herald of Industry* is the only paper in the

United States entirely edited by women and the mechanical work entirely executed by girls, excepting the presswork.

"The editor of this journal leads a busy life. She teaches type-setting, physical culture, reform dress-making, sets up the advertisements, edits, reads proof, and makes up the paper, invents and makes her own dresses, bonnets, etc., does her own housework, canvasses for the *Herald* and keeps up a large correspondence. She claims the 'Woman Worker's Champion Belt' — who will contest it?"[35]

A noteworthy manifestation of her interest in printing occurred at the 1883 Mechanics' Institute's annual exposition in San Francisco. She not only exhibited "works of Sisters of Social Science," but also samples of "wood and metal types, and printers' implements; copies of newspapers, etc."[36]

Before and after the 1884 campaign for national office, Marietta L. Beers-Stow's newspaper was issued as *National Equal Rights* with Washington, D. C., added to San Francisco as a place of publication. It apparently ceased to exist after the February 1885 issue. Following the campaign, she listed herself as "authoress and publisher" and then "journalist" in the San Francisco Directory and, finally, as "widow." She wrote and spoke widely, her principal means of participation in the political events of the day after the demise of her newspaper, but her influence waned once she had stepped down. When Belva Lockwood again ran for president in 1888, Marietta Stow was no longer on the ticket. As with Emily Pitts Stevens and her strong feminist views, Mrs. Stow was the subject of many articles in San Francisco newspapers, many of which spoofed the themes of her dress and dietary reforms.

After her prominence in political and self-betterment matters, Marietta Beers-Stow devoted all her time to philanthropic work and lectured widely. She also founded "what she was pleased to call the Birdie Bell Republic in Alden [Alameda County] and there sought to instruct children in methods of government by making them self-governing." Her interest in women printers apparently waned completely after 1885 and no longer did typesetting give "more pleasure — not excepting love-making."

Following two attempts at suicide because of a severe illness — "her sufferings were intense" — Marietta Beers-Stow died on 27 December 1902 at the age of 64 and was cremated.[37]

Fig. 39: *Belva Lockwood and Marietta Stow, first woman candidates for president and vice president of the United States on a national ticket. From* Woman's Herald of Industry. Courtesy California State Library.

Chapter 9   Investigation and Revelation

Julia Schlesinger, the "outspoken, radical and reformatory [her words]" editor of *The Carrier Dove*, always kept an alert eye on her exchanges and acquired most of her suffrage and other feminine news from these sources. Although one article she printed is heavy with irony and may contain localisms, it does give insight into the customs and working conditions some woman typesetters endured: "Of all the occupations in which a woman can engage for the purpose of making a living the most thankless is that of setting type, says *The Denver Tribune*. The female compositor leads a weary and dreary life. She is never permitted to strike a *fat take*, she is denied the inestimable boon of setting up the thoughtful matter which emanates from the editorial room; she is never reckoned capable of handling manuscript [typewriting was common by 1889], and the very idea of her being competent to set up a display head is deemed atrocious. She is expected to hammer away at miscellaneous reprint; the only [typesetting] bonanza she ever strikes is solid minion [7 pt.] with an occasional oasis of leaded brevier [8 pt; less work to get to 1,000 ems] when the business manager concludes that advertising is dull enough to admit of the biggest kind of type. But this is not all — no, the worst remains to be told. When the work is done for the day, it is not with the female printer as with others of the trade. She can not adjourn to a convenient and comfortable saloon and play pedro [a card game] or old sledge [seven-up, a card game] for the beer or throw dice for five-cent cigars or jeff [a dice-like game using quadrats] for the drinks. She must pick her way home through all sorts of weather to a dreary room and cold bed. She has no wife to thrash, no children to scold, no furniture to break — none of those sweet luxuries which are supposed to be part and parcel of the glorious art preservative [printing]. As a class, female printers are diligent and worthy. They never 'sojer' [finesse for a choice of *takes*], they never bother the editors for chewing tobacco; they never prowl around the exchanges for the *Police Gazette* [a girlie newspaper]; they never get themselves full of budge [booze] and try to clean out rival print shops; they never swear about the business manager; they do not smoke nasty old clay pipes, they never strike for more pay, they do not allude to editorial matter as 'slush' or 'frogwash;' in short, they are patient, gentle, conscientious and reliable. They peg [set] right along for $7 a week, dress tidily, keep solid with the foreman, and, last of all, when the female compositor gets tired of her treadmill, unceasing round of toils, she marries the best looking printer in the shop, and then she becomes a verier slave than before."[1]

Under the revised International Typographical Union (ITU) rules of 1876, the locals had full discretion concerning the admission of women, but San Francisco Local #21 still was not interested in them. The appearance of Mrs. Louise M. Wheeler by transfer from Washoe (Nevada) #65 in June of 1876 must have been both a surprise to all and an irritation to many. Under the rules, San Francisco #21 was compelled to accept her transfer: "This is the first lady admitted into the San Francisco local and makes quite a departure from the old rule."[2] There is no record of her actually working as a typesetter in San Francisco.

Seven years later, the pressure had grown so great and the locals had been so inconsistent under an optional plan, the ITU opened membership to women in all locals in 1883. This move not only made the better-paying work available to females but also put them officially on a parity in wages — at least in Union shops. Once admitted, the Union's defense of women was notable and the woman typesetters were grateful for the protection and higher wages the Union commanded for its members. The reason there were comparatively few females in Union positions may have been suggested unwittingly by Amanda Slocum: only the best and most reliable were chosen by both the Union and employers, and records confirm that Union women generally remained on the job, year after year, married or not. In short, the Union women represented the ablest and most stable of the female typesetters, and some had the best-paying jobs among women with this skill, meaning newspaper work.

However, the Union never gave up trying to exert complete control, so a sudden strike hit the *Call* and *Bulletin* on 30 July 1883 because these plants also employed non-Union men. The next day, the *Call* said: "Among those who very kindly and cheerfully hastened to assist the *Call* out of what appeared to be a dilemma [the strike was called at 6:30 p.m.] were many ladies. . . . One lady who had been connected with the *Call* composing room for years, and who was the mother of four children, hearing of the threatened embarrassment [the *Call* was a morning newspaper], tendered her services, and cheerfully wielded the 'stick' and 'rule' throughout the night and performed her self-allotted task with as much dispatch, cheerfulness and ability as the gentlemen who had just abandoned them could possibly have done."[3] The *Bulletin* had no trouble getting the typesetters they needed either, and the strike failed completely. It was truly ironical that women again assisted in breaking a strike but this time they were also opposing female Union members, something unimaginable in the recent past.

The matter of labor generally had now become so important that the California State Legislature authorized a new Bureau of Labor Statistics (BLS), also in 1883, about the same time the federal Department of Labor was established to succeed the Bureau of Labor. The reports of both had a striking similarity with a heavy emphasis on women in all trades, especially printing, and both afforded many insights and details not previously available.

The third California BLS report noted that the 1880 Census showed 14,142 women as wage earners in San Francisco and estimated that with the city having grown about 50% in population by 1888, 20,000 would not be an underestimation by then. They were working in more than three hundred occupations.[4] A survey of general conditions in ten "printing houses" revealed variances in pay and working conditions that had probably been typical for many years. In one instance, fourteen females were employed from 7:30 a.m. to 5:30 p.m. with a half hour for lunch. "Wages from 25 to 30 cents per one thousand ems. Workroom clean but not well ventilated; it is close to a market, the odors from which are offensive and injurious; there is but one water-closet, which is in a filthy condition."

The second place visited was the optimum in its ambiance if not the pay: "Seven females employed, from 18 to 24 years of age. Scale of prices 30 cents per one thousand ems. Workroom on top floor, clean, good light, and well ventilated; separate water-closets." A firm had twenty women at 30 cents per thousand ems and the "Workroom [was] crowded, well lit but not clean." In one shop, the report said, "water-closets very dirty; washing facilities bad" and also described another's water-closet as "very filthy" with no washing facilities.[5] There apparently were good reasons, other than money, for women to change jobs frequently.

Of special interest is the section on the home conditions

of working women, separated by trade. Thirty-six females in printing were surveyed. One lived with her mother, the most frequent circumstance, but she had other means of support and gave all her wages to her mother. She had been through grammar school and was "well dressed." Another only worked for "pin money" while a third "is well educated; lives in nice house owned by father; only works for pocket money and to be independent; buys her own clothes." Included was a graduate of a normal school who lived with a widowed mother who owned her home. Virtually all the women reported giving their wages to their mothers or to the family to assist with living expenses. An unusual example was a woman living with her mother-in-law. Her husband supported her "entirely" while she worked "to pass time" and "saves all her wages."[6] In view of the need for so many women to work, these kinds of responses are surprising.

The report also observed: "The supply of female wage earners bears a far greater proportion to the demand than in the case of males. This tends to keep women's wages down." Charts showed this over and over again for women in printing. The highest weekly wage for women was $17.00, the lowest $3.00 with an average of $6.00 for "compositors," a misuse of the term in this context. The hours of work were now down to nine.[7]

Both the federal and state reports for 1888 surveyed prostitutes to discover their previous occupations. San Francisco produced 441 replies (56 refused), but none reporting said they were former typesetters. This was noteworthy in view of the range of work divulged: music teachers, telegraphers, a translator of Spanish and an actress who had been married. It also reflected the better pay that women received for typesetting, even at the low end of the scale, than for other occupations. The federal report surveyed 3,866 prostitutes nationwide (statistics for San Francisco were curiously missing) with no typesetters either but revealed hairdressers (1,155) and "no previous occupation" (1,236) as the principal former call-ings. The range ran from "feather curlers" to "telegraph and telephone operators."[8]

In the BLS report, a table showing trade-union rosters revealed the tremendous growth in San Francisco in printing. Local #21 of the ITU had 630 members and there were 245 in Pressman's Local #24, which by this time would have included women press-feeders. In an analysis of Local #21, the BLS said: "There are perhaps two hundred and fifty competent printers outside the union, besides about one hundred and fifty girls. Counting every one who has work in a printing establish-ment, there are over six hundred men, girls, and boys in this city outside the union. The union has a fixed scale of wages; it ranges on day work from $18 to $30 a week. Job and book printers are paid from $18 to $25 a week for time work. On piece work a man [or woman] can make about $18 a week, if he [or she] works full time. The positions on the daily newspaper are paid somewhat higher ($5 a day of ten hours, with seven composing and three distributing, but the work is very ex-haustive, and a man cannot continue for more than a few days). . . .The number of female printers in the job printing es-tablishments who are members of the union is remarkably small — only three out of a total of fifty-one, but this number has much increased since the investigation by the bureau [ear-lier that year, see below]." Noteworthy is one assertion (p. 348): "As a general rule, apprentices receive no wages for the first three months; after that period they are paid from $1 to $3 per week," a practice that undoubtedly had been in effect for a very long time and was a continuing source of much friction with the Union. Many apprentices and the resultant low wag-es enabled an employer to underbid almost anyone and thus take much-needed work from the higher-paying shops.[9]

Two preliminary statistics are of interest. First, the WCPU was shown as having five male and one female Union employees, one non-Union male and one boy and six girl apprentices at about the time Mrs. L. G. Richmond & Son was incorporated into The Hicks-Judd Company. Second, A. J. Lafontaine, established in 1858, had one woman and one man working in 1883, each non-Union. A rare business Directory of 1885 shows Mrs. Lafontaine as owning the firm and this is the only instance during the period of this study of a woman succeeding to her husband's business.[10]

Two companies, in particular, were notorious for their exploitation of women and youth and became such a problem that the Union and some competing firms made representation to the Commissioner of Labor Statistics who responded with a special investigative hearing. Its focus was the status of women in printing.[11] Primarily, the investigation was aimed at Bacon & Co. and a very large, church-run enterprise across the Bay in Oakland, Pacific Press Publishing House. Jacob Bacon (1834–1895), a native of Maine, had been around San Francisco since the 1850s, sometimes with a partner but in later years by himself. He took advantage of the cheap labor that women typesetters afforded as time went on, although as noted above he apparently paid them the same scale as men in earlier days. As the century moved forward and competition became keener, he saw the employment of women and youth as a means to outbid his competition and always had an imbalance of apprentices, who were paid nothing or the very lowest scale. Said Commissioner Tobin: "Mr. Bacon . . . received my deputy . . . very curtly, and refused to give the number of apprentices employed, or the wages paid to his employees. He said the number of women and girls employed was twenty. My agent learned afterwards from Mr. Bacon's forewoman, that the number was only fourteen, of whom four were apprentices. The rates of wages paid were given to the latter as follows: 25 cents per 1,000 ems for composition on newspapers; 30 cents for 1,000 ems for composition on books.

"Apprentices are bound by contract or indenture to serve a term of four years. For the first three months they receive no wages; for the next three months they are paid $12 per month, and for each succeeding six months they receive $3 per month advance. Ten per cent of this amount is retained by the principal [Bacon] as a guarantee for the fulfillment of the contract, and in case the apprentice should leave, or be discharged for cause, before the expiration of the term of apprenticeship, the whole amount retained is forfeited. No interest is allowed on the wages kept back by Bacon & Co."[12] The Commissioner concluded that "This contract or indenture is a remarkable document, on account of its one-sidedness and demonstrates the necessity of having a good apprenticeship law on our books."[13]

The Pacific Press Publishing House became one of the largest employers of women from the time of its founding in 1876 (18 are listed in Appendix A). The plant began with a "four-roller cylinder press" and was thus equipped to do large jobs from the outset. Business grew rapidly with a corresponding need for added space and equipment. Accordingly, in 1880, "another press (a Cottrell & Babcock four-roller, double-revolution), the largest book press on the coast, was imported and put in operation. Another book press was added in 1882 as well as additional machinery in the book-bindery and electrotype foundry." The original building also had been expanded to accommodate the new requirements and new structures erected.[14]

From owning the water supply for the steam boilers, the source of power for the large presses, to importing paper, designing, engraving, electrotyping and stereotyping, this was a complete and self-sufficient plant, probably, as they claimed, the most elaborate in the West. It was also aggressively competitive and many books and periodicals bear its imprint as a result. Pacific Press was established as an adjunct to the Seventh Day Adventist Church and produced all its printing as well. It also maintained an office in San Francisco because the city was a major source of its work. The BLS revealed that by 1883 Pacific Press Publishing House had a capital investment of $48,000 and a twelve-month output worth $78,000 at a labor cost of $34,500. During the previous twelve months, the firm had employed a maximum of seventy-eight persons at one time, with an average of forty-two males and twenty-four females. The telling differential came with the rates of pay. Skilled women received $1.25 a day and the men $2.50; women received $1.00 a day for "ordinary labor."

When the hearing adjourned to Oakland, Commissioner Tobin sent "my lady assistant" to visit the boarding house run by Pacific Press for employees at 714 Eleventh St. "It contains sixteen rooms, with parlor. The rooms are light, airy and comfortably furnished. Dining room and kitchen are on ground floor. The charge per week for board is $3. Room rent ranges from $4 to $7.

"The rules regulating this boarding-house are very stringent, as the boy and girl boarders are committed to the care of the Matron, who looks after their moral welfare. I was assured that the boarding-house was not a paying investment, and it was sustained by the association solely for the accommodation and well being of their employees with whom it is optional to avail of its advantages."[15]

All testimony for both hearings was given under oath and began in San Francisco. John Duckel, a Union printer whose wife, Alice (A 69), was a typesetter, claimed that the system of working for three months without pay only applied to girls at Bacon & Co., the men and boys being paid from the start. Frank O'Neil found Bacon "overbearing." Miss Mamie Sweeney (A 265) testified that she was paid 45¢ per 1,000 ems, averaging weekly $18 to $20. "I would advise all girls to join the union, and hope never to sever my connection with it."

Miss Grace Ellis (A 76) noted that Bacon "invariably measured the work of all girls employed on newspaper work," where typesetting by 1,000 ems was crucial. Another woman said, "Mr. Bacon had treated her boorishly," while a married woman testified, "Mr. Bacon's treatment of the girls was rough and indelicate." Ms. Frances Auld (A 8) apparently learned to set type at the WCPU, and Bacon wanted her to work for nothing despite a year's experience. When she did go to work for him, she said: "I did not like it at all. The place is poorly ventilated, and scarcely a day passed but what I suffered with a severe headache. . . . Mr. Bacon had a method of giving the best work to his old employees and the poorest to the new hands [i.e., "so-jering" on his own]." This comment was in marked contrast with that of a man who reported, "In a union office women have to take work from the same hook [as the men]."[16]

An especially revealing comment was made by another male: "I have heard some men remark some ill feeling of girls or women working in an office with them; they are called the 'bum' element; the object of the union is in opposition to discrimination." James H. Barry, a master-printer whose firm exists to the present time, said Bacon "treated me badly, and in fact every one of his employees."

Some interesting testimony about Bacon never appeared in the printed report. In a story headlined, "Female Compositors, Bacon & Co.; Partial to Those of Pretty Form and Face," the *Examiner* said: "On the reporter's table was some choice written testimony of lady compositors that was not intended to be given away . . . . It had been obtained by a lady in the employ of the Commissioner who had made a still-hunt expedition to the non-union house of Bacon & Co. It was racy with blush-producing testimony that a member of the firm was partial to pretty-faced, plump-figured young ladies for copyholders in his establishment, where he read the proofs."[17]

Commissioner Tobin's summation of the San Francisco investigation concluded: "The . . . testimony clearly demonstrates the advantages to be derived from organization. Female compositors [typesetters], not members of the Typographical Union, earn from $8 to $10 per week, while members earn about double that amount. . . . All bore witness to the kindly feeling which existed between the men and women of the Typographical Union."[18]

When the hearing moved to Oakland for the problems at Pacific Press, virtually all the testimony was about wages. A point made by one woman is especially noteworthy: she "never had any portion of my wages deducted for religious or other purposes." (The Seventh Day Adventists have always been a tithing religious organization.) The manager of Pacific Press gave a lengthy statement which included this remark: "The reason why girls are taught the trade in one year, while boys take three years is that girls devote themselves to but one department [typesetting], while boys work in several branches [p. 19]." In his summation, the Commissioner noted: "In the case of proof-readers men receive precisely double the wages of the women. . . . A person unversed in Adventist methods would imagine that such would be the very reason she should receive more. In the other departments, women receive about three fifths of the wages paid to men."[19]

Pacific Press is a long-lived business and as late as 1978 "another discrimination case against Pacific Press Association" was filed. The plant had a plan of different wages for different kinds of work plus a policy of paying heads of households more whatever the job. This was declared to be discriminatory towards married women who worked and so a suit was brought.[20] In 1993 — one-hundred-ten years after the Tobin investigation — another report revealed women still make less than men and have "fewer opportunities and decision-making responsibilities than their male counterparts . . . . Women make only 75¢ for every dollar a man makes, even when they both have the same job and identical qualifications."[21] Unfortunately, the 1883 BLS report might have been written today for many industries and jobs; the battle for equality in the workplace is far from won.

The leading national trade journal published an interesting suggestion regarding health problems. "We believe also that the establishment of a gymnasium, with bathrooms attached, by the larger typographical unions, would prove a paying investment in more than one sense, and be the means of removing many of the nervous diseases now caused by high pressure work and foul atmosphere."[22] Typesetting has never been a "romantic" occupation, especially in the generally unsatisfactory working conditions typical of the labor-intensive 19th Century.[23]

By 1890, the woman's movements had begun to stir positively again in San Francisco, but none of them was involved with printers or female publications as in the past. Once Marietta Beers-Stow had ceased to function with her school for typesetters and her newspaper, no one ever succeeded her. Only one woman, details on whom are unfortunately lacking, attempted to start a printing business on her own. Jennie Patrick (A 202) appeared as a book and job printer in San Francisco, but living in Oakland. The very next year, she was listed as working at the "Herald of Trade Publishing Co.," in 1890 as "solicitor" for them and in 1893, 1894 as a printer for the same company which advertised their book and job shop. Two billheads are the only known surviving examples of her personal enterprise, "Jennie E. Patrick, prnt, 413–415 Market St., 1st Floor."[24]

For many years, inventors had been trying to create a practical and economical typesetting machine both to speed up the process and to reduce the cost of labor. Mark Twain, a former printer himself, lost a fortune on such a device — this is why he did so much lecturing in later years; he needed the money. Ottmar Mergenthaler (1854–1899), an immigrant watchmaker from Germany, conceived an altogether different concept in 1884: matrices (dies) were assembled in a typographical line using a keyboard and then cast as a single entity, hence the name "Linotype." The *New York Tribune* first used this revolutionary machine successfully in 1886, and from that time the linotype gradually succeeded in dominating newspaper and book work. "It took about twenty years for the linotype and its mechanical associates [intertype, a rival of linotype, and monotype, which cast and set one piece of type at a time] practically to supplant straight-matter typesetting by hand. And although *one* man at a linotype came to achieve as much as *four* men had done by hand, the widening use of printing kept the larger proportion of compositors at work."[26] Again, equating the words compositor and typesetter gives a misleading impression of the situation. If women typesetters were to continue in the composing room, additional training would have been required both for standard work of compositors or the new machine at an increase in wages. The already-trained compositors, who were paid by time, might have been able to keep their jobs in an expanding printing economy, but woman typesetters were not able to retain their lower-level jobs — already few enough in number — which were being phased out entirely by the new typesetting processes.

According to one excellent study, the Union added its own complications for its policy was "based on the requirement that the machines should be operated only by journeymen printers," that is, those who had completed a full apprenticeship, which women had rarely done, even after they were able to join the Union. There was an important "economic advantage" from the introduction of mechanical typesetting: "the material reduction secured in the length of the working day. The number of machines employed by 1901 nationwide was 4,138 in newspaper offices and 837 in book and job offices. . . . Allowing for time spent in distribution of type and in pasting up 'dupes' [for measurement], the usual working day on newspapers was rarely less than ten hours." As might have been expected, "No tendency to replace male with female labor has ever appeared. The proportion of female to male operators is smaller than the proportion of female to male hand compositors. In January 1904, the number of women operating typesetting and typecasting machines in the United States

171

and Canada was 520, about 5% of the total number of linotype operators. The number of women engaged in the United States in 1900 as printers and compositors was 15,875, about 15% of the total number of printers and compositors.[26] A history of the San Francisco Union notes that "The decade of the nineties was one of depression succeeded by prosperity . . . . Faced with unemployment . . . the union was forced to cope as well with the introduction of [typesetting] machinery in the trade." By 1894, a "machine scale committee" established the Union's principles: "piece work on machines was prohibited;" an eight-hour day was to be the new standard; and "members of the union 'who will be most affected by the introduction of the machines' were to be given the opportunity to operate them."[27] Based on the record at that time, this training policy could have included women, but the record is silent and the likelihood is that the Union initially trained only journeymen, i.e., those who previously had undergone a full apprenticeship.

The 1890 Census was completely destroyed by accident, so no industrial figures for San Francisco or Northern California exist. With women already greatly in the minority in the trade before the linotype, there undoubtedly were fewer of them after the linotype came more generally into use. So, in a cyclical sense, women at the end of the century were about back where they started. It would not be until the 1950s and 1960s when photo-typography and photo-offset methods of printing began to ease out the linotypes and monotypes and traditional letterpress that women finally found many more opportunities than had ever existed in the past for them in the printing business, including presswork. The advent and increasing use of personal computers from 1980 on expanded their opportunities greatly. A new kind of enterprise, the "instant printing" establishment, has opened other jobs in labor and management for women. Important, too, is the fact that working conditions have changed dramatically. In the San Francisco Bay Area, Union printers work 35 hours a week, for example, and the lighting and general ambiance of the work-place generally have no resemblance whatever to the negative conditions that Commissioner Tobin reported.

There is a large, neat building on a frontage road near San Francisco Bay which proclaims in huge, bold, sans-serif letters, "Joyce Printing Company" and beneath, in large script lettering, "Susan Joyce, Owner." It is a pleasing example of the positive presence of women in printing in Northern California today.

Fig. 40, following page: *Enclosed within this cover were three other pieces, including the menu.* Author's Collection.

SONGS

FOR THE

FIRST ANNUAL DINNER

OF THE

Yale Club of the Pacific Coast

— AT —

The Baldwin

December 13th, 1877.

Women's Print, 424 Montgomery St., S. F.

NOTES

CHAPTER 1

1. Joseph Moxon, *Mechanick Exercises on the Whole Art of Printing*, ed. Herbert Davis and Harry Carter (London: Oxford University Press, 1958), p. 12.
2. Aulus Flaccus Persius, *Satyrae* (Paris: Badii Asencii, 1523). Asencius was also the first to use a printing press in his device.
3. Daniel Berkeley Updike, *Printing Types, Their History, Forms and Use*, 2 vols. (Cambridge: Harvard University Press, 1937), 1:9n.
4. Geoffrey A. Glaister, *Glaister's Glossary of the Book* (Berkeley & Los Angeles: University of California Press, 1979), p.13:1–2.
5. Ruth Shepard Graniss, "Printer Maids, Wives and Widows," in *Bookmaking on the Distaff Side* (New York: [The Distaff Side], 1937), p. [5]. This book consists of a series of separately printed, unpaged signatures.
6. Colin Clair, *A Chronology of Printing* (London: Cassel & Co., Ltd., 1969), p. 26:1.
7. Luther J. Ringwalt, comp., *American Encyclopedia of Printing* (Philadelphia: Menamin & Ringwalt, 1871), p.190:2.
8. There is a copy in the Mills College Library. I am indebted to Librarian Emeritus Flora Elizabeth Reynolds for this reference.
9. Virgil Solis, *Effigies Regum Francorum Omnium…* (Nuremberg: [In officina typographica Katherinae Theodorici Gerlachii relictae viduae, et haeredum Johannis Montani]), 1576. This book is also noteworthy for its Jost Amman illustrations.
10. Henry R. Plomer, *A Dictionary of Booksellers and Printers … in England … 1641–1667* (London: The Bibliographical Society, 1907), p. 168. I am indebted to Ms. Barbara Tandy, my former student, for this reference.

11. Graniss, "Printer Maids," p. [6].
12. Leon Voet, *The Golden Compasses*, 2 vols. (Amsterdam & New York: Van Gendt & Abner Schram, 1969, 1972), v. 2:331.
13. See reference to Jacob Bacon, San Francisco, on this subject in Chapter 9.
14. Voet, *Compasses*, v. 1:143–4.
15. Graniss, "Printer Maids," [p. 6], notes further that "the five Plantin daughters who grew up in the business, serving as copy-holders before they were twelve years old…must have enjoyed the work, for they all married employees of their father… ."
16. Thomas W. McDonald, *Obelisk* (Los Angeles: Thomas W. M'Donald at Black Mack the Handpress, 1961), p. 6.
17. *The Autobiography of Benjamin Franklin*, ed. Leonard W. Labarre et al. (New Haven: Yale University Press, 1964), p. 145.
18. Ibid., p. 166.
19. Isaiah Thomas, *The History of Printing in America*, 2 vols. [1874] (New York: Burt Franklin, 1964), v. 1:195.
20. Lawrence C. Wroth, *The Colonial Printer* (Charlottesville: The University Press of Virginia, 1964), p. 155–156.
21. James J. Brenton, "Female Printers," in *Voices from the Press* (New York: Charles B. Norton, 1850), p. 302.
22. Leona M. Hudak, *Early American Women Printers and Publishers, 1639–1820* (Metuchen, New Jersey: The Scarecrow Press, 1978), p. 677.
23. Ibid., p. 4.
24. Ava Baron, *Woman's "Place" in Capitalist Productions, A Study of Class Relations in the Nineteenth Century Newspaper Printing Industry*, Ph.D. dissertation, New York University, 1981 (Ann Arbor: University Microfilms International, 1981), p. 107. This provocative study integrates much valuable information and many useful references despite the fact that the act of setting type by hand is described as follows: "While reading the manuscript the compositor stood over the case [sic] containing the type characters sorted into order [sic]. With a pair of tongs [!] he selected a character at a time and assembled a line of print [sic] on [sic] the composing stick… ." (pp. 83–84). See Chapter 2.
25. Isaiah Thomas, *History*, v. 1:358.
26. Helen L. Sumner, *History of Women in Industry in the United States, Vol. IX, Bureau of Labor Report on Condition of Woman and Child Wage-Earners in the United States* (Washington: Government Printing Office, 1910), p. 212.
27. Baron, *Woman's "Place,"* pp. 174–175.
28. Sumner, *Women in Industry*, p. 215. Typesetting machines were never an issue in San Francisco from 1857–1890. All the problems — and more — discussed here are fully covered by Sumner.
29. George E. Barnett, "The Printers, A Study in American Trade Unionism," *American Economic Association Quarterly*, October 1909, p. 316. The whole issue (387 pp.) is devoted to the perceptive and accurate descrip-

tion of composing room methods and trade relationships, both practically and historically.

30. John Southward, *A Dictionary of Typography and Its Accessory Arts* (London: Joseph M. Powell, 1870–1871), passim.

31. Ringwalt, *Encyclopedia*, passim.

CHAPTER 2

1. Thomas MacKellar, *The American Printer: A Manual of Typography*, 5th ed. (Philadelphia: MacKellar, Smiths & Jordan, 1870), p. xii.

2. An American scholar-printer, Theodore Low DeVinne, first advanced these conclusions in *The Invention of Printing* (New York: Francis Hart & Co., 1876; reprint, Detroit: Gale Research, 1969), pp. 66–67. This is still a useful book despite the great amount of research since its first printing.

3. Richard N. Schwab, et al., "Cyclotron Analysis of the Ink in the 42-line Bible," *The Papers of the Bibliographical Society of America*, Vol. 77, No. 3, 1983, pp. 285–315.

4. Helmut Presser, "Johannes Gutenberg: To the 500th Anniversary of the Death of the Inventor of the Art of Printing," *Export Polygraph International*, January 1968, p. 8.

5. For a detailed discussion of the one- and two-pull press, see: Michael Pollak, "Performance of the Wooden Printing Press," *Library Quarterly*, April 1972, pp. 220–227.

6. Leon Voet, "The Making of Books in the Renaissance as Told by the Archives of the Plantin-Moretus Museum," *Printing & Graphic Arts*, December 1965, p. 34.

7. "Making the Magazine," *Harper's New Monthly Magazine*, December 1865, p. 7:1.

8. The "w," a double "v," was another latecomer, but found its place within the standard lay of the case during the transition period leading to the creation of the single-character "w."

9. Luther Ringwalt, comp., *American Encyclopedia of Printing* (Philadelphia: Menamin & Ringwalt, 1871), p. 118:1. The author's copy has the rubber stamp of "Frank Eastman & Co., printers, April 7, 1881, 509 Clay St., S. F."

10. Ibid., p. 118:2. "The relative stature of the workmen can also be not inaccurately determined by the different heights of their frames," according to a composing-room visitor in Great Britain, in 1839: Sir Francis Bond Head, *Printers, Poor Devils* (Berkeley, California: A. R. Tommasini, 1961), p. 5. I am indebted to printer Bruce Washbish for reminding me of this reference.

11. Thomas Lynch, *The Printer's Manual, A Practical Guide for Compositors and Pressmen* (Cincinnati: Cincinnati Type Foundry, 1859), p. 64.

12. MacKellar, *American Printer*, p. 109.

13. Head, *Printers*, p. 12.
14. A. N. Sherman, *The Printer's Manual* (New York: West & Trow, Printers, 1834), p. 11. This manual is noteworthy for showing 109 different *imposing schemes*, including one for a 20-page *signature*.
15. George Trumbull, *Pocket Typographia: A Brief Practical Guide to the Art of Printing* (Albany: George Davidson, 1846), p. 92. A copy of this rare manual is in the Robert Grabhorn Collection at the San Francisco Public Library.
16. Lead is a cumulative poison. "Printing offices, even the cleanest and brightest and most welcome in the world are unhealthy. ... Lead dust and oxides of lead carried in the air, shaken from type cases and tables in use, lifted from the floor while walking, falling into open receptacles for drinking water ... taken into the mouth with drinking water or from the fingers while eating lunch, breathed into the lungs, absorbed through the pores of the skin, afflict most printers with plumbaic [lead] poisoning ... ." *Printers and Printing in Providence* (Providence, Rhode Island: Providence Typographical Union, 1908), p. 180. The only other extended reference to composing-room health found appeared under a heading of "Consumption Among Compositors:" "The position in which the compositor is required to stand, the irregularity of the hours of labor, the insufficient ventilation of the workrooms, and the general want of cleanliness, are acknowledged to be the chief factors of evil." (*Inland Printer*, April 1884, p. 10:2.) The remainder of this article discusses ventilation problems at length and the eating of food without washing hands after setting type, etc.
17. The term "stick" remains because they were originally made of wood. By the mid-19th Century, wooden composing sticks were made only for setting wood types.
18. The common denominator of the em is 60 which affords 25 mathematically related combinations for the spaces normally accompanying a given font/size of type. See Hugo Jahn, *Hand Composition, A Treatise on the Trade and Practice of the Compositor and Printer* (New York: John Wiley & Sons, Inc., 1931), p. 143, for a chart showing all 25 possibilities.
19. International Typographical Union, *Bureau of Education, ITU Lessons in Printing, Elements of Composition — Unit 1* (Indianapolis: International Typographical Union, 1942), Lesson 5, p. 17.
20. Ringwalt, *Encyclopedia*, pp 116:2–117:1, shows nine varieties of composing sticks, the only difference being the manner in which the slide is adjusted to the desired measure, i.e., by a screw, locking lever, knee, etc.
21. Jobbing — job — work was ordinarily placed in a smaller galley, usually measuring 8 1/4" x 12 3/4".
22. Joseph Moxon, *Mechanick Exercises on the Whole Art of Printing*, ed. Herbert Davis and Harry Carter, 2nd ed. (New York: Dover Publications, Inc., 1978).
23. "Corpus Typographicum, A Review of English and American Printers'

Manuals," in Lawrence C. Wroth, *Typographic Heritage* (New York: The Typophiles, 1949), pp. [56]–[90].

24. Moxon, *Mechanick Exercises*, p. 328.

25. John Southward, *A Dictionary of Typography and Its Accessory Arts* (London: Joseph M. Powell, 1871), p. 8:2, "Companionships." There is no record of companionships in the West, although they were common in the East early in the 19th Century.

26. Ringwalt, *Encyclopedia*, p. 253:1, who also describes the procedure. Moxon says that "To play at quadrats . . . either for money or for drink" is subject to a solace (fine) and then describes the game in detail (*Mechanick Exercises*, p. 324). The secretary of the San Francisco Typographical Union reported in 1889 that business was so bad that the newspapers had "an average 'jeff' of eleven cases on each paper per day . . . ." (*The Typographical Journal*, 15 September 1889, p. 6:3.) That is, there were more typesetters than were needed so whether one worked on a given day literally depended on chance. Union records also show that there were women typesetters at almost all the newspapers.

27. Trumbull, *Typographia*, p. 34. Typographer Dan X. Solo says that "working the hook" is a synonymous phrase of an indeterminate date.

28. *Pacific Printer, Stationer and Lithographer*, September 1883, p. 282:1. Horace Greeley's copy was widely thought to be the worst of all, and stories abound concerning who really "wrote" what editorial.

29. Ringwalt, *Encyclopedia*, p. 190:1, shows a *copy-holder*, also called a *guide*.

30. *Woman's Pacific Coast Journal*, May 1871, p. 8:3. The first eight issues were printed by the Women's Co-operative Printing Union.

31. Based on an average of 6,500 ems per 10-hour day, at 75¢, this equals $4.88 while wages for typesetting by women in the Towne & Bacon Records show a maximum of $3.00 per day. This differential would have allowed for the markup in the billing.

32. O'Meara & Painter Records, Kemble Collections, California Historical Society.

33. *Printers and Printing in Providence*, p.117

34. *Investigation by the Commissioner of the State Bureau of Labor Statistics into the Condition of Labor in Printing Establishments of San Francisco and Oakland, February and March, 1888* (Sacramento: State [Printing] Office, 1888), p. 9.

35. *San Francisco Examiner*, 4 March 1888, p. 2:2. With this issue, the *Examiner* became the property of William Randolph Hearst (p. 10:2).

36. MacKellar, *American Printer*, 12th ed., 1879, p. 227, shows such a ruler; see also, Ringwalt, *Encyclopedia*, p. 478, who calls it a "triangular type-gauge."

37. Ringwalt, *Encyclopedia*, p. 304:1.

38. Thomas Lynch, *The Printer's Manual* (Cincinnati: The Cincinnati Type Foundry, 1859), pp. 236–237.

39. The "Standard of Type" of the Washoe (Nevada) Typographical Union Local #65, for example, took into account the alphabet length of differ-

ent sizes of text types. "The following scale of measurement is adopted by this Union: Pica [12-pt.], 12 ems to the alphabet; Small Pica [11-pt.], 12; Long Primer [10-pt.], 12; Bourgeois [9-pt.], 12; Brevier [8-pt.], 13; Minion [7-pt.], 13; Nonpareil [6-pt.], 14; Agate [5-pt.], 15; Pearl [4 1/2 pt.], 16; Diamond [4-pt.], 17. Fonts of type falling below this scale, shall be cast up [measured] according to the width of type, by the rule of three, thus: If the alphabet measures 10 ems in width, as 10 is to 12, so is 1,000 to 1,200; and the font shall always be counted thus: Every 1,000 ems as 1,200; if the alphabet of a font of type shall measure 11 ems in width, as 11 is to 12, so is 1,100 to 1,200; the font counting every 1,000 ems as 1,100." Also, "No work to be measured in a type larger than Pica [12-pt.]." Armstrong, *Nevada Printing History . . . 1858–1880*, p. 396. Sacramento Typographical Union #46 had similar rules, but a marked-up copy for the 1882 edition in the author's possession eliminates the rule of three. The San Francisco Typographical Union #23 had no provisions for alphabet length in 1885, according to its scale of prices (Bancroft Library).

40. ITU, *Composition, Lesson 10*, pp. 4–5.

41. Theodore Low De Vinne, *The Practice of Typography . . . Plain Printing Types* (New York: The Century Co., 1902), p.113.

42. Ibid., p. 118.

43. Theodore Low De Vinne, *The Printers' Price List, a Manual for the Use of Clerks and Book-Keepers in Job Printing Offices* (New York: Francis Hart & Co., 1871), p. 425. This quotation describes fat typesetting exactly. De Vinne's Printers' Price List went into several editions and was the standard for a large number of printing-offices for many years.

44. George E. Barnett, "The Printers," *American Economic Association Quarterly*, October 1909, p. 550.

45. Nevin is saying that the appearance — *layout* — of job work was a function of the compositor and not marked up in advance of her taking the copy from the hook.

46. The "office towel" is another item understood by all but not recorded in printers' manuals or standard references.

47. San Francisco *Morning Call*, 20 October 1892, p. 6:3. This article is verbatim except for making longer paragraphs out of the short ones common in the narrow columns of newspapers. For consistency, one set of quotation marks has been moved to conform to prior usage in the story. I am indebted to Allan R. Ottley for finding this reference.

48. Charles Munsell, comp., *A Collection of Songs of The American Press and Other Poems Relating to The Art of Printing* (Albany: [Joel Munsell], 1868), p. 123.

CHAPTER 3

1. Robert E. Riegel, *American Feminists* (Lawrence & London: The University

Press of Kansas, 1963), p. 9. The first American edition of *A Vindication* also appeared in 1792; there were also other editions dated before 1800.

2. Eleanor Flexner, *Century of Struggle* (Cambridge, Massachusetts: The Belknap Press of Harvard University Press, 1975), p. 77.

3. Riegel, *American Feminists*, pp. 49–50.

4. Flexner, *Century*, pp. 76, 77. Forty of the three hundred attendees were men.

5. Harriet Martineau, *Society in America*, 2 vols. (London: Saunders & Otley, 1837), v. 2:131–151. By 1870 in San Francisco, these were still the principal occupations open to women with box-factory work substituting for the non-existent cotton mills.

6. Miriam Gurko, *The Ladies of Seneca Falls; The Birth of the Woman's Rights Movement* (New York: Schocken Books, 1976), p. 26.

7. Ibid., p. 28.

8. W. J. Rorabaugh, *The Craft Apprentice* (New York: Oxford University Press, 1986), p. x.

9. An explanation of printing-office terminology and practices needed for understanding this study appears passim in Chapter 2.

10. Helen L. Sumner, *Report on Women and Child Wage-Earners in the United States*, Vol. IX, *History of Women in Industry in the United States* (Washington: Government Printing Office, 1910), p. 258.

11. Ava Baron, *Woman's "Place" in Capitalist Production: A Study of Class Relations in the Nineteenth Century Printing Industry*, Ph.D. dissertation, New York University, 1981 (Ann Arbor: University Microfilms International, 1981), p. 158. Unfortunately, the "West" means only as far as the Mississippi River to the author. However, her discussion of printing-office practices, in both newspaper and book and job offices, is very useful (pp. 160–175).

12. James J. Kenneally, *Women and American Trade Unions* (St. Albans, Vermont: Eden Press Women's Publications, Inc. 1978), p. 4. This book offers an excellent short history of the many involvements of woman organizations' leaders in the labor movements from 1865–1875.

13. Gurko, *The Ladies*, p. 29.

14. Joan M. Jenson and Gloria Ricci Lothrop, *California Women: A History* (San Francisco: Boyd & Fraser Publishing Co., 1987), p. 65.

15. George A. Stevens, *History of Typographical Union #6* (Albany: J. B. Lyon Co., 1913), p. 422.

16. *Boston Traveler*, 20 December 1849. I am indebted to Ms. Mary Silloway, a doctoral candidate in the School of Library and Information Studies, University of California, Berkeley, for this citation. See also John W. Tebbel, *A History of Book Publishing in the United States*, 4 vols. (New York & London: R. R. Bowker, 1972), v. 1:407–408.

17. *Printer's Circular*, October 1868, p. 233:1, 2. This report goes into great detail while claiming other specific failures of the women.

18. International Typographical Union, *A Study of the History of the International*

Typographical Union, 1852-1963 (Colorado Springs: The Executive Council, International Typographical Union, 1964), pp. 236, 239. (See Chapter 4 of this study for the development of an earlier Female Typographical Union in San Francisco in 1864 which had no connection with the ITU.)

19. Gunther Barth, "Emergent Urban Society in the Far West" (Paper delivered at a Department of History Colloquium of the University of California at Berkeley, 3 May 1966), p. 1.

20. James D. Hart, comp., *A Companion to California*, 2nd ed. (Berkeley, Los Angeles & London: University of California Press, 1987), p. 437:1.

21. Ray Allen Billington and Martin Ridge, *Westward Expansion, A History of the American Frontier*, 5th ed. (New York: Macmillan Publishing Co., Inc., 1982) p. 2.

22. Roger W. Lochtin, *San Francisco, 1846-1856, from Hamlet to City* (New York: Oxford University Press, 1974), p. vii.

23. Mason, Jack and Helen Van Cleave Park, *Early Marin*, 2nd rev. ed. (San Rafael: Marin County Historical Society, 1976), p. 69, n9.

24. Ann Braude, *Radical Spirits, Spiritualism and Women's Rights in Nineteenth-Century America* (Boston: Beacon Press, 1989), p. 2. This is a most informative and useful study.

25. Elizabeth Cady Stanton, Susan B. Anthony, Matilda Josyln Gates, eds., *History of Woman Suffrage*, 4 vols. (Rochester: Fowler & Wells, 1881–1902), v. 3:530.

26. Ann Braude. "News from the Spirit World . . . ," *Proceedings of the American Antiquarian Society*, vol. 99, part 2, p. 405.

27. R. Laurence Moore, *In Search of White Crows* (New York: Oxford University Press, 1977), p. 39. I am indebted to Michael Rosen for this and several other Spiritualism references.

28. Braude, *Radical Spirits*, pp. 57, 58.

29. See, for example, *Alta California*, 17 December 1852, p. 1:7.

30. John T. Bonnel, *Phenomena of Spirit Manifestations . . . Also Directions for Forming Circles and Ascertaining Who Are Mediums* (San Francisco: Alta California Steam Printing Establishment, 1852).

31. Julia Schlesinger, *Workers in the Vineyard* (San Francisco: The Author, 1896), p. 26:2. Relevant publications will be described as they appear in the sequence of this study.

32. Braude, "Spirit World," p. 407.

33. Braude, *Radical Spirits*, p. 192.

34. Ibid, p. 152.

35. Jane Rendall, *The Origins of Modern Feminism: Women in Britain, France and the United States, 1780-1860* (New York: Shocken Books, 1984), pp. 257–258.

36. *The San Francisco Alta California for the Atlantic States and Europe*, 30 September 1854, p. 7:1.

37. *San Francisco Daily California Chronicle*, 30 September 1854, p. 2:2.

38. Ibid., 2 October 1854, p. 2:1.

39. *Daily Alta California*, 2 October 1854, p. 2:4.
40. Edward C. Kemble, *A History of California Newspapers*, ed. Helen Harding Bretnor (Los Gatos, California: The Talisman Press, 1962), pp. 223, 260, 331.
41. *Alta California*, 17 April 1880, p. 1:2.
42. *The San Francisco Golden Era*, 1 October 1854, p. 2:2.
43. *The Yreka Ladies' Budget*, 13 February 1856, p. [2]:1. There is a copy in The Bancroft Library.
44. Ibid., p. [4]:3
45. Kemble, *California Newspapers*, p. 204. A copy of the first issue is in The Bancroft Library.
46. Receipt Book, 1857, Whitton & Towne, Vol. 26 of the Towne & Bacon Records, Stanford University Library. I am indebted to Librarian Bruce Johnson for directing my attention to the women in these files.
47. All identifiable Whitton & Towne and Towne & Bacon woman employees are listed in Appendix A. Early San Francisco printing-offices sometimes included bindery operations of varying capacities and women did indeed work in this part of the business, as shown in Census and Directory entries. The records cited here give no indication about bindery work by its employees.
48. *The Hesperian*, 5 May 1858, p. 8:3.
49. Ibid., 15 June 1858, p. 61:2.
50. Ibid., 15 November 1858, p. 216:1.
51. *San Francisco Chronicle*, 19 October 1990, Section C, p. 20:1.
52. *Constitution and By-Laws of the San Francisco "Ladies' Protective and Relief Society"* (San Francisco: Whitton, Towne & Co., Printers, 1853), pp. [3], 4.
53. The California Labor and Employment Exchange organized in the early part of 1867 and from the start of operations on 27 April 1868 to 1 July 1870 placed 24,580 persons, 6,581 of whom were women. Its credo was to find work "for all applicants, irrespective of sex or nationality." Unfortunately, many of the female placements were in domestic service "in the interior" and similar traditional woman's work. "No fees were exacted." (*Langley's San Francisco Directory, 1871*, p. 45.)
54. *Constitution, By-Laws, and Act of Incorporation of the Ladies' Union Beneficial Society of San Francisco* (San Francisco: Printed by B. F. Sterett, 1861), p. [3]. The incorporation date was February, 1861.
55. Peter R. Decker, *Fortunes and Failures, White Collar Mobility in Nineteenth-Century San Francisco* (Cambridge, Massachusetts: Harvard University Press, 1978), p. 211.
56. Ping Chiu, *Chinese Labor in California, 1850–1880, An Economic Study* (Madison: The State Historical Society of Wisconsin, 1967), p. 58.
57. John S. Hittell, *The Commerce and Industries of the Pacific Coast* (San Francisco: A. L. Bancroft, 1882), p. 645.
58. Ibid., p. 645.

59. Jenson and Lathrop, *California Women*, p. 37.
60. *The San Francisco Union Printer*, November 1888, p. 3:1.
61. *Union Printer*, February 1889, p. 4. Two of the eight Union newspaper shops had no women typesetters.
62. *Hesperian*, September 1860, p. 336.
63. *Report of the Second Industrial Exhibition of the Mechanics' Institute* (San Francisco: Frank Eastman, Printer, 1859), p. 87.
64. *San Francisco Business Directory and Mercantile Guide* (San Francisco: B. F. Stilwell & Co., 1864), p. 249.

CHAPTER 4

1. Robert Chandler, "A Woman Printer Battles the All-Male Union," *The Californians*, March/April 1986, p. 45:2.
2. Ruth Shaw Worthing, "Sophia Emeline Walker," ms. version (without cuts) of "A Woman Wielding Words" (*NEWMONTH*, October 1984, pp. 13–17), pp 1–9. I am indebted to Historian Chandler for a copy of the Worthing ms. as well as several other references below.
3. According to *Webster's International Dictionary*, a columbiad was "A heavy, long-chambered, muzzle-loading gun...designed for throwing shells and shot at a high angle of elevation."
4. *San Jose Mercury*, 28 July 1864, p. 2:3.
5. *Pacific Monthly*, December 1863, p. 332. *The Hesperian* changed its name in June 1863.
6. Ibid., February 1864, pp. 437–438. The many typographical errors and the gross misuse of commas have been corrected in transcribing this quotation.
7. David F. Selvin, "History of the San Francisco Typographical Union" (M.A. diss., University of California, Berkeley, 1936), pp. 8–34.
8. *Daily Morning Call*, 3 February 1864, p. 2:1.
9. Ibid., 14 February 1864, p. 8:4.
10. *Pacific Monthly*, March 1864, p. 498. The use of quotes around Lisle Lester indicate an awareness that this was a pseudonym.
11. Ibid.
12. Ibid.
13. Payroll records of Whitton & Towne and Towne & Bacon, at Stanford University, include the names of women employees in 1857, 1858 and beyond who, by the nature of their rate of pay and other indications, were undoubtedly typesetters. See Special Note 3, Appendix A.
14. *Pacific Monthly*, March 1864, p. 499.
15. *Napa Register*, 3 March 1864, p. 2:1.
16. *Pacific Monthly*, May 1864, p. 622.
17. *The Virginia* [Nevada] *Evening Bulletin* (29 April 1864, p. 3:3) asked, "What

business have the male bipeds in a female organization? Are they long-tailed gentry?" I am indebted to Mr. Michael Rosen for this note.

18. *Watsonville Pajaro Times*, 7 May 1864, p. 2:2.

19. To assure a record, women from Sacramento, Santa Cruz and Merced are included in Appendix A, although they are outside the primary areas of this study.

20. Most of the errors occur in "The Editor's Table"; the larger portion of any given issue was set to the standards of the time.

21. *Pacific Monthly*, June 1864, p. 673.

22. Ibid., August 1864, p. 825.

23. Ibid., July 1864, p. 745.

24. There were thriving coal mines in Contra Costa County at this time but her employee might have left for the gold country, a more common occurrence. The foreman was probably George Sprague.

25. Lisle Lester was "publisher" of the magazine; "printer" might have been more accurate in this context.

26. *Pacific Monthly*, July 1864, p. 746.

27. *Daily Alta California*, 23 July 1864, p. 1:1.

28. *Langley's San Francisco City Directory, 1871*, p. 868:1.

29. *Pacific Monthly*, June 1864, p. 683.

30. This information was printed in the corresponding issues of *Pacific Monthly*.

31. Journals of Towne & Bacon, vol. 9, p. 6, dated 6 April 1864. These Journals show that Towne & Bacon was heavily engaged in "trade work," that is, composition or presswork for other printers, both locally and out of town. The Journals are in Special Collections, Stanford University.

32. Ibid., p. 26. Unfortunately, there is no clue whether the posters were for one of her reading tours or for the promotion of her magazine.

33. Ibid., 11 May 1864, p. 44. Why Lisle Lester was charged only $18.00 as opposed to $33.00 for the April issue might be explained by the inclusion of composition or binding in the cost for the magazine's production.

34. Ibid., 22 July 1864, p. 129.

35. Eureka Printing House (San Francisco) billhead, The Kemble Collections, California Historical Society.

36. She had said only that the magazine "circulated to two thousand readers" in her advertisement in *San Francisco Business Directory and Mercantile Guide* (San Francisco: B. F. Stilwell & Co., 1864), p. 249.

37. *Pacific Monthly*, October 1864, p. 817 (a verso).

38. Ibid, November 1864, p. 886.

39. *Grass Valley Daily Union*, 16 April 1865, p. 3:1.

40. *Sacramento Daily Union*, 16 September 1867, p. 4:1.

41. Worthing, "Sophia Walker," p. 12.

42. *Alpine Miner*, 30 November 1867, p. 3:1.

43. *San Jose Weekly Mercury*, 27 July 1871, p. 2:4.

44. *Chronicle*, 13 August 1871, p. 3:4.
45. Letter to Lyman Draper, head of the Wisconsin Historical Society, 5 February 1877.
46. Fond du Lac (WI) newspaper obituary, 25 June 1888.

CHAPTER 5

1. Henry Eno, *Twenty Years on the Pacific Slope*, ed. W. Turrentine Jackson (New Haven and London: Yale University Press, 1965), p. 88.
2. *Alpine Chronicle*, quoted in the *Sacramento Union*, 25 January 1868, p. 2:3. The 1870 Census lists no woman typesetters in Alpine County, which then had a population of 870. There is the possibility that the women cited may have only worked part time and thus were not enumerated as typesetters.
3. San Francisco *Daily Alta California*, 1 February 1868, p. 2:2.
4. San Francisco *Daily Examiner*, 9 July 1868, p. 3:3. The *Examiner*. as was customary, only listed cabin passengers. There were 730 in second-class and steerage on the ship.
5. *The Revolution*, 10 September 1868, p. 149:2. The letter was dated 14 August 1868. The first woman applied formally for admission to the Union in November 1868.
6. San Francisco *Daily Evening Bulletin*, 6 August 1868, p. 3:3.
7. 1868 was the date of publication of the *Directory for 1869*.
8. *Bulletin*, 6 August 1868, p. 3:3.
9. *San Francisco News-Letter and California Advertiser*, 8 August 1868, p. 9:1.
10. Henry G. Langley, comp., *San Francisco Directory* (San Francisco: Henry G. Langley, Publisher, 1868), bottom half of insert opposite p. 560.
11. *Los Angeles Star*, 22 August 1868, p. 2:1. The whole editorial article on the subject covers nine column inches.
12. *The Revolution*, 10 September 1868, p. 149:2. The week before, Mrs. Peterson ran an ad in *The Revolution* for her new printing-office and signed herself as "Business Manager" (p. 143:2).
13. *San Jose Patriot*, 14 August 1868, p. 2:1.
14. *San Bernadino Guardian*, 13 September 1868, p. [2]:5.
15. Langley, *1868 Directory*, p. 896. Other incorporation data have been found in *Directories* and the California State Archives. See also, Chapter 6.
16. *California Weekly*, 5 May 1868, p. 5:2.
17. Langley, *1871 Directory*, p. 45. Significantly, the Laborers' Eight-Hour League organized in September 1868, had 500 members (Langley, *1869 Directory*, p. 841:2).
18. *San Francisco Banner of Progress*, 1 December 1867, p. 2:5.
19. Ibid., 29 February 1868, p. 2:3.
20. *Daily Alta California*, 18 January 1869, p. 1:2.

21. *El Dorado*, 19 February 1869, p. 4:1, 2.
22. *Elevator*, 5 March 1869, p. 2:3. Afro-Americans published this newspaper.
23. *Los Angeles Weekly News*, 13 March 1869, p. 2:2.
24. *El Dorado*, p. 4:1.
25. Ibid., 20 March 1869, p. 4:1.
26. Ibid., p. 8:4.
27. The file in The Bancroft Library ends with the September 1869 issue.
28. Advertisement, *The Traveler's Guide*, July 1869. The Wells Fargo History Museum is presently in a new building at this address. See Appendix B for a checklist of imprints.
29. *Saturday Evening Mercury*, 22 May 1869, p. 1:2.
30. Ibid, 12 June 1869, p. 1:2. The complete prospectus, dated 5 June 1869, was also printed.
31. Incorporation papers, California State Archives.
32. Before the Stevens-Wickes takeover, the *California Weekly Mercury* said that for each woman to get what she felt was rightfully hers, " . . . she works with others to produce a common fund, which is divided weekly, and which will furnish her a larger return than if she worked alone. . . . [Women's Co-operative Union] members are not day-laborers, but joint-stockholders working on their own account, and at wages agreed upon in council by themselves. . . . There are already one hundred and fifty working members, who will be able to earn, on an average, about $45 per month; of this, twenty per cent will be retained to pay the loan [to start the enterprise], and the expenses of the institution." (15 March 1868, p. 4:2.) Female typesetters could earn at least double the amount quoted here. There is no evidence the WCPU ever operated in the manner cited.
33. "Of many ways to put co-operation in practice it is well to refer here to only one — the stock company. Employers who wish to try co-operation will find this an excellent way to experiment." (*The Pacific Printer, Stationer and Lithographer*, February 1884, p. 5:2.)
34. Incorporation papers, California State Archives.
35. Henry R. Wagner, "Commercial Printers of San Francisco from 1851 to 1880," *Papers of the Bibliographical Society of America*, v. 33, 1939, pp. 69–84.

CHAPTER 6

1. Under an earlier name, *California Weekly Mercury*, the paper noted, "Talented Lady writers are constantly employed in efforts that lend grace to our columns "(11 June 1865, p. 5).
2. Frances E. Willard and Mary A. Livermore, eds., *A Woman of the Century . . . Leading American Women in All Walks of Life* (Buffalo, Chicago, New York: Charles Wells Moulton, 1893), p. 686:1–2.

3. Emily Pitts Stevens never hyphenated her name and her preference will be the style used here. The misspelling of Pitts as "Pitt" was common and some modern researchers unfortunately have perpetuated both errors.

4. California State Board of Health, Certificate of Death #2132.

5. *The Revolution*, 10 August 1871, p. 2:1. This description and other information strongly suggest that Susan B. Anthony and Emily A. Pitts were acquainted before Miss Pitts moved to San Francisco.

6. G. P. R. James (1801[?]–1860) was a prolific writer of historical romances. A story in *The Revolution* said Ms. Pitts came to San Francisco "in company with the *daughters* [emphasis supplied] of G. P. R. James (10 August 1871, p. 2:1).

7. The *Directory* listings for Charles Miehl's educational enterprise vary. In 1864, he was named "professor and principal, Young Ladies' Seminary (and Kindergarten), south side of Geary between Stockton and Powell;" his residence is given as 41 South Park. The following year, the name had changed to French and English Female Institute and occupied 54 and 55 South Park where it is listed in the *1866 Directory* but gone subsequently. By 1867, Emily Pitts is listed as "teacher, Calisthenics Home Institute," and residing at 122 Taylor Street. Florence James appears as "assistant, Home Institute," and living at the same address. There is no further record of the Calisthenics Home Institute but this may have been a name given by the two women to their private endeavors.

8. *Municipal Reports for the Year 1866-67* (San Francisco: Board of Supervisors, 1867), pp. 398, 400; Ibid., 1867–1868, p. 434.

9. *Pioneer*, 15 October 1870, p. 1:2. I

10. *Revolution*, 10 August 1871, p. 2:1.

11. *Sunday Mercury*, 24 January 1869, p. 2:2.

12. Ibid., 31 January 1869, p. 2:1.

13. Ibid., 7 February 1869, p. 2:2.

14. Ibid., 28 February, 1869, p. 2:2.

15. Ibid., 18 April 1869, p. 2:3. The standard workday at this time was 10 hours. The eight-hour law passed by the legislature was to take effect in April 1868 after the governor signed the bill (San Francisco *Daily Morning Call*, 15 February 1868, p. 1:3).

16. Ibid., 25 April 1869, p. 2:2.

17. *Saturday Evening Mercury*, 2 May 1869, p. 2:2. This was the name of the newspaper when it first appeared in 1865.

18. Ibid., 5 June 1869, p.1:2.

19. Ibid., 14 August 1869, p. 1:3.

20. Ibid., 18 September 1869, p. 2:2.

21. Ibid., 31 August 1869, p. 2:2.

22. *Revolution*, 7 October 1869, p. 220:1.

23. *Pioneer*, 13 November 1869, p. 1:2.

24. Elizabeth Cady Stanton et al., *History of Woman Suffrage*, vol. 3, p. 752. The new group was inspired by the National Woman Suffrage Association which was founded in New York, May 1869, with "Mrs. Elizabeth T. Schenck as vice-president for California."

25. The Wells Fargo Museum is currently at 420 Montgomery Street.

26. *Great Register of the County of San Francisco, 1869*, p. 375. He had originally registered to vote 21 July 1866, at age 22.

27. *A Woman of the Century*, p. 686:1. As noted above, her name is correctly Mrs. Emily Pitts Stevens, although many different sources, ranging from newspapers to books, give it as "Pitt Stevens," and a hyphen appears passim in many different references. Her name never appeared with a hyphen in the masthead or text of *The Pioneer*. The *1871 Directory* misprinted husband Augustus's address as 530 Greenwich Street.

28. Daughters of the American Revolution, *Vital Records from the San Francisco Bulletin* (N.p.: California State Society, 1870, 1871, 1872). There is no record in other San Francisco newspapers checked yet this was a period when marriages in Winnemucca, Nevada, Grass Valley, California, and other outlying towns were regularly listed so there is the possibility that no marriage ever took place. *A Woman of the Century*, p. 686:1, gives her marriage date as 1871, possibly a misprint. The details and length of this entry, which includes a photograph, suggest some information may have been supplied by the subject.

29. *Pioneer*, 5 January 1871, p. 1:5.

30. "Petition for Woman Suffrage," in *Appendix to Journals of Senate and Assembly, of the Eighteenth Session of the Legislature of the State of California*, vol. 2 (Sacramento: D. W. Gelwicks, State Printer, 1870), pp. 3–34.

31. The name is misprinted as "J. N. Choyuski" in the petition.

32. *Pioneer*, 19 March 1870, p. 1:1.

33. Ibid., 20 August 1870, p. 1:2.

34. *Pacific Union Printer*, November 1888, p. 3:2–3.

35. Ibid., 27 August 1870, p. 1:3.

36. *Pioneer*, 23 March 1870, p. 1:1.

37. *Workingman's Journal*, 24 September 1870, quoted in *The Pioneer*, 1 October 1870, p. 4:7.

38. The trial and acquittal of Laura Fair for the murder of her lover was a San Francisco sensation of this time and became central to the question of "Free Love" that perversely affected the woman's movement.

39. *Daily Morning Call*, 19 May 1871, p. 3:4.

40. Ibid., 18 May 1871, p. 3:2. Total expenses of the two-day gathering were $180, income was $124.45. The deficit of $35.55 was picked up by Mrs. Eunice S. Sleeper. John A. Collins was president of the Association at this time.

41. *Daily Morning Chronicle*, 10 July 1871.

42. Beverly Beeton and G. Thomas Edwards, "Susan B. Anthony's Woman

Suffrage Crusade in the American West," *Journal of the West*, April 1982, p. 7:1.

43. *Call*, 12 July 1871, p. 3:3. The *Call* printed the complete texts of the three public lectures.

44. *San Francisco Bulletin*, 26 May 1871, p. 3:5.

45. *Call*, 13 July 1871, p. 3:3.

46. *Daily Alta California*, 13 July 1871, p. 1:2. As early as 20 January 1867, the *Alta* had said, "We cannot discern what interest of society would be compromised by the votes of women" (p. 2:1).

47. Mrs. Mary G. Snow was the activist-wife of Herman Snow, "spiritual bookseller and agent for Adams & Co.'s golden pens" (*1872 Directory*).

48. *Call*, 15 July 1871, p. 3:3.

49. *Pioneer*, 25 December 1869, p. 1:2.

50. *Oakland Daily Transcript*, 13 March 1872, p. 2:2.

51. *Sacramento Bee*, 14 March 1872, p. 3:2.

52. *Chronicle*, 15 March 1872, p. 1:1.

53. Incorporation papers, California State Archives #10866.

54. *Pioneer*, 19 December 1872, p. 6:4. A later version of this ad was used both in *The Pioneer* and in *The University Echo*, a University of California student publication.

55. Henry R. Wagner, "Commercial Printers of San Francisco from 1851 to 1880," *Papers of the Bibliographical Society of America*, vol. 33, 1938, pp. 69–84.

56. *Pioneer*, 13 August 1870, p. 1:4. For a brief history of Train, see *Dictionary of American Biography*, ed. Dumas Malone (New York: Charles Scribner's Sons, 1936), v. xviii: 626–627. He was not only a contributor to *The Revolution*, but had also given the funds to found the newspaper.

57. Ibid., 15 August 1872, p. 4:1.

58. Ibid., 29 August 1872, p. 4:1.

59. *Chronicle*, 2 June 1872, p. 1:1.

60. *Pioneer*, 20 June 1872, p. 5:2. In a story about an evening session (21 June 1872, p. 3:3), the *Chronicle* said, "Mrs. Stevens occupied the chair with a gorgeous pair of green ear-rings in her ears."

61. *Daily Examiner*, 21 June 1872, p. 3:2.

62. Ibid., 22 June 1872, p. 3:2.

63. Yerba Buena is now principally a connective for the San Francisco-Oakland Bay Bridge and man-made Treasure Island is also linked to it.

64. *Examiner*, 24 June 1872, p. 2:4.

65. Mrs. Loomis and Mrs. Olmsted, active suffragists, were listed as widows in the *1872 Directory*.

66. *Bulletin*, 26 June 1872, p. 3:4.

67. *Call*, 25 June 1872, p. 3:6. The story nearly filled a column.

68. *Examiner*, 26 June 1872, p. 3:4.

69. *Pioneer*, 2 January 1873, p. 4:1.

70. *Chronicle*, 10 April 1873, p. 3:5.

71. Ibid.
72. Ibid., 11 April 1873, p. 3:5.
73. Ibid.
74. Ibid., 27 April 1873, p. 5:3– 5.
75. Ibid., p. 5:4.
76. *Alta*, 7 May 1872, p 1:2.
77. Ibid.
78. *Pioneer*, 1 October 1873, p. 2:2. The newspaper apparently did not survive long thereafter.
79. *Common Sense*, 7 November 1874, p. 303:1.
80. Elizabeth Putnam Gordon, *Women Torch Bearers*, 2nd ed. (Evanston, Illinois: National Women's Christian Temperance Union Publishing House, 1924), p. 318.
81. *Call*, 1 November 1891, p. 12:3.
82. Ibid., 6 February 1895, p. 5:2. The previous year, Mrs. Stevens had told one newspaper, "If there is one thing I long for it is the enfranchisement of women. I not only hope, but believe I will vote for the next President of the United States. I have been lobbying in a quiet way for some time and I find the men favorable to the suffragist bill. . . ." (*Examiner*, 20 December 1894, p. 20:5–6.) This was a prelude to the suffragists again going to Sacramento to lobby the Legislature.
83. San Francisco County Clerk's Office: Conservator's and Appraiser's Reports File #12944.
84. California State Board of Health, Certificate of Death #2132.
85. *Chronicle*, 14 September 1906, p. 15:7.

CHAPTER 7

1. For a listing, as well as other woman imprints, see Appendix B. Virtually every job bore an imprint which also contributed to the firm's high visibility.
2. *1871 Directory*, p. 844:1; also *1872 Directory*, p. 907:1, the last time such data were given.
3. *Business Directory of San Francisco and Principal Towns of California and Nevada, 1877* (San Francisco: L. M. McKenney, 1877), p. 582. The *1877* (SF) *Directory* named her as "Proprietress" and the *1879 Directory* listed her firm as L. G. Richmond & Son for the first time. The charter was not forfeited until 13 December 1905 "for failure to pay the license tax for the year ending 30 June 1906" (California State Archives, Certificate of Incorporation #10888).
4. Author's interview with the Heatherly Family. Her descendants, of whatever generation, referred to her as "Grandma" and had always believed she was a binder (her second husband was a principal in the lead-

ing bindery of its time). It was news to them that she had been a capable and prominent master-printer.

5. Dorothy N. Spear, "City Directories of the United States through 1869," *Bibliography of American Directories*, (Microfiche), pp. 169, 180, 188.

6. Benjamin F. Wilbour, comp., *Little Compton Families* (Little Compton, Rhode Island: Little Compton Historical Society, 1967), p. 525.

7. Over the years, some *Directories* called this street "Williams," but William is correct. The Richmond home was not many yards away from the modern-day location of Capwell's department store. William St. is at an angle to Telegraph Avenue.

8. "[I]n a folio each sheet has been folded once, in quarto twice, in an octavo three times; the size being thus respectively a half, a quarter and an eighth of the original sheet [approximately 19" x 25" in this instance]." John Carter, *ABC For Book Collectors* (New York: Alfred A. Knopf, 1978), p. 100. The inside-chase measurements in these presses were: 12" x 18", 9" x 14" and 5.5" x 10" respectively.

9. *San Jose Weekly Mercury*, 11 November 1869, p. 1:5.

10. Charles A. Murdock, "History of Printing in San Francisco," *Pacific Printer and Publisher*, May 1925, p. 365:2.

11. *San Jose Mercury*, 11 November 1869, p. 1:5.

12. Ibid., 4 August 1870, p. 2:3. The *Call* and *Bulletin* were under the same management at this time; the *Alta* and the *Chronicle* were paying 75¢ at the start of the strike.

13. San Francisco *Morning Call*, 6 August 1870, p. 2:3.

14. Ibid., 14 August 1870, p. 2:3.

15. *San Francisco Bulletin*, 15 November 1870, p. 3:2.

16. *San Jose Mercury*, 3 November 1870, p. 3:1.

17. *Sacramento Daily Union*, 17 December 1870, p. 4:7.

18. The business was on the second floor.

19. This was below the community standard for men both before and after the 1870 strike.

20. *San Jose Mercury*, 22 December 1870, p. 2:4. The article appeared as "Our San Francisco Letter" and was signed, "Breton." Later, Editor Owen acquired a faltering daily's equipment in San Jose, went daily himself and commented: "We think we may safely say that we now have one of the snuggest, cosiest and most convenient printing offices to be found in the State . . . . In the distribution of our force, we have five day hands, three of whom are ladies — all accurate and rapid compositors. They 'set up' the miscellany and such matters as can be set by daylight." In other words, for safety's sake no women were employed on the all-important night shift. (Ibid., 21 March 1872, p. 2:6.)

21. Jacob Bacon was pilloried by the Union for this practice. Interestingly enough, when he finally met the Union's terms and became a Union shop, he had more women employees than men, the only non-female

managed business with this imbalance.

22. 12 May 1879; Heatherly Family Papers.

23. The author acknowledged she published her memoir to raise money for herself. This book was reprinted in 1992 by The Book Club of California for its members.

24. *San Francisco* (Bishop) *1875 Directory*, p. 13B. Printing in a foreign language creates many special demands, especially in the training of typesetters to hyphenate properly, to recognize language peculiarities and in proofreading. The printing-office also must have an adequate supply of accented characters in all type sizes, another sizable investment.

25. The author's WCPU ephemera collection reflects this variety.

26. *Woman's Pacific Coast Journal*, August 1870, p. 58:3.

27. Ibid., December 1870, p. 125:3.

28. *San Francisco Post* was consistently interested in the cost of living and ran stories which give an insight into the purchasing power of known wages. There was every type of accommodation from first-class hotel rooms costing $3+ to rooming houses charging 25¢ to 50¢ cents a night. Boarding-houses were common in a wide range of prices. Many addresses in Appendix reflect those establishments and also suggest many of them lived at home. Because of the tiring, 10-hour day in force during this period, the appeal of a boarding-house is understandable. Apartments, as we know them today, were mostly used by families at this time. Many woman typesetters apparently had no desire to return from work to cook. If they earned $20 to $25 per week, they should have been fairly comfortable. Their board must have been within their ability to pay because the prices of commodities were reasonable. Apples were 10¢ to 20¢ per half peck; oranges were 25¢ to 40¢ a dozen; a quart of strawberries cost 15¢; meats averaged 13¢ a pound; cheese 22¢; salt 2¢; mackerel, 25¢ each; chicken averaged 6.5¢ per pound and vegetables averaged about 12¢ a pound. *Post*, 7 May 1871, p. 1:2–3; 23 July 1871, p. 1:5–6.

29. *San Jose Mercury*, 5 January 1871, p. 2:1.

30. *Woman's Journal*, February 1871, p. 159:2. A large amount of research has failed to reveal anything about "Oriental" women having worked in this printing-office. The firm had been owned in 1869 and 1870 by Max Weiss who often used the word Oriental in his imprint. The probability is that the word did not necessarily apply to the employees.

31. *Report of the Seventh Industrial Exhibition of the Mechanics' Institute* (San Francisco: Women's Co-operative Print), p. 60. There were three paper mills and three type foundries in operation at this time.

32. *Bulletin*, 4 February 1870, p 3:6. At least one local foundry was capable of making steam locomotives.

33. *Pacific Printer*, January 1878, p. 3:2. No copy is presently known. This was a publication of the Miller & Richard (Scotland) type foundry's San Francisco branch office.

34. *San Francisco Examiner*, 12 November 1875, p. 3:8.

35. Blake, Robins Ledger E, Kemble Collections, California Historical Society.

36. A book in progress will require much more paper than many small jobs.

37. *Farley's Reference Directory of Booksellers, Stationers and Printers in the United States and Canada* (Philadelphia: A. C. Farley & Co., 1885).

38. *Printers Guide*, December 1877, p. 2:2; January 1878, p. 2:2.

39. *Pacific Printer, Stationer and Lithographer*, May 1884, p. 11:1. "These presses can be run by hand at a speed of 1,000 an hour, and by steam up to 1,500 an hour." January 1881, p. 4:1.

40. *Carrier Dove*, April 1887, p. 147:1.

41. Ibid., 25 February 1888, p. 139:3.

42. Ibid., 11 August 1888, p. 520:2–3; 18 August 1888, p. 535:1.

43. *Pacific Printer and Publisher*, February 1937, p. 46:3. The firm name became Hicks & Judd in 1880. Judd died 11 January 1937.

44. 1876–1878 Marriage Records, Alameda County Recorder's Office, Book E, p. 240.

45. *Investigation by the Commissioner of the State Bureau of Labor Statistics into the Condition of Labor in Printing Establishments of San Francisco and Oakland, February and March, 1888* (Sacramento: State [Printing] Office, 1888), p. 4.

46. California State Archives.

47. One Women's Co-operative Printing Office billhead in the author's collection is rubber-stamped in red, "The Hicks–Judd Co., successors to," while a later The Hicks–Judd Co. billhead has this same information neatly printed in a panel.

48. *Call*, 4 September 1901, p. 12:1. The *Call* also said the building was owned by the Regents of the University of California. The *Chronicle* noted the roof fell in on the bindery and composing room on the third floor and that there was water damage to the pressroom (4 September 1901, p. 12:5–6).

49. *Pacific Printer*, February 1937, p. 46:3.

50. *Call*, 16 October 1898, p. 15:3.

51. *Oakland Tribune*, 15 October 1898, p. 2:4.

52. *Oakland Enquirer*, 11 October 1898, p. 5:3. Willard was possibly a source for this story.

53. Certificate of Death Record, Alameda County: *Index to Deaths, Book "D,"* pp. 336–337.

54. The California State Library, Sacramento; The Bancroft Library, UC Berkeley; The Huntington Library, San Marino; and the North Baker Library, California Historical Society, San Francisco, all have significant holdings of WCPU imprints.

CHAPTER 8

1. San Francisco *Morning Call*, 18 May 1871, p. 3:2.
2. *1860 Census*, Santa Clara County, p. 465.
3. Charles E. Slocum, *History of the Slocums ... of America ...*, 2 vols. (Defiance, Ohio: Published by the Author, 1908), v. 2:260.
4. Pay Roll Records, Vol. 26.
5. *History of the Slocums*, v. 2:260.
6. Truman J. Spencer, *The History of Amateur Journalism* (New York: The Fossils, Inc., 1957), p.162.
7. *The Olive Branch*, October 1873, p. 2:1
8. *Santa Cruz News*, 24 August 1859, p. 4:3. This was vol 1, #1.
9. *Common Sense*, 5 June 1876, p. 1
10. W. N. Slocum, *Revolution* (San Francisco: [The Author], 1878), p. 17.
11. Ibid., p. 24.
12. *Golden Gate*, 20 February 1886, p. 5:2. This publication was edited by J. J. Owen.
13. *Common Sense*, 16 May 1874, p. 8. The paper has stood the test of time well and is not deteriorating as are so many 19th Century papers. The S. P. Taylor & Co. paper mill was the first in California.
14. Ibid., 25 July 1874, p. 123; 7 June 1874, p. 45.
15. It is interesting to note that no political zealots were officers or directors, mostly persons with a business background and one a female physician.
16. *Common Sense*, 20 June 1874, p. 68; 18 July 1874, p. 116.
17. Ibid., 27 March 1875, p. 536; 13 February 1875, p. 464.
18. Incorporation Certificate #10857, California State Archives. Here, too, the officers and directors were principally from the world of business rather than political advocacy. This was an ambitious undertaking: "Capital shall be $100,000 divided into twenty thousand shares of the par value of Five Dollars each."
19. *Common Sense*, 3 April 1875, p. 551.
20. Ibid., 5 June 1875, p. 25.
21. Ibid, p. 27.
22. Stanton et al., *History of Woman Suffrage*, v. 3:761–762. Clara married and became Clara Eldridge while working for her mother. She died prematurely in November of 1880.
23. It is worth noting that Amanda Slocum did not work at typesetting or any allied chore but only in the business office of Taylor & Nevin. This suggests the possibility she never may have worked as a hands-on person in a printing-office. An intriguing question arises: did Howard P. Taylor, a San Francisco printer since 1856, bear any relationship to her first husband?
24. Annagret Ogden, "Marietta L. Stow: 19th-Century Candidate for Gov-

ernor and Vice President," *The Californians*, May/June 1987, pp. 6–7, 67. See also: Mary Virginia Fox, *Lady for the Defense* (New York and London: Harcourt Brace Jovanovich, 1975), p. 135. This book is a biography in the form of a novel.

25. The UPI sent out a story when Geraldine Ferraro was running for vice president listing women's candidacies for national office which acknowledged the pioneering of the Lockwood-Stow ticket ("Women in Politics," *Santa Barbara News-Press*, 13 July 1984, p. A 5).

26. Edward T. James et al., eds., 2 vols., *Notable American Women, 1607–1650* (Cambridge: The Belknap Press of Harvard University Press, 1971), v. 2, p. 415:1

27. *National Equal Rights*, November 1884, p. 4:3, p. 5:2. This was the name of the monthly newspaper founded by Mrs. Stow, during the presidential election campaign. The original name still appeared in the running titles on the inside pages.

28. *Lady for the Defense*, p. 134.

29. *Woman's Herald of Industry*, September 1881, passim. The leaf size of this newspaper is 13" x 18".

30. Ibid., January 1882, p. 2:1;

31. Ibid., February 1882, p. 5:3.

32. Hawks, of Alameda, California, is extremely important in the history of typefounding for he is credited with the American Point System and initiating the standardization of type measurement in the U.S.A. and Great Britain.

33. *Woman's Herald*, February 1882, p. 1:3.

34. Ibid., February 1882, p. 1:3; October 1882, p. 5:3.

35. Ibid., April 1883, p. 1:20; May 1883, p. 1:3; July 1883, p. 4:4; September 1883, p. 1:3.

36. *Report of the Seventeenth International Exposition of the Mechanics' Institute of the City of San Francisco* (San Francisco, Mechanics' Institute, 1883).

37. *San Francisco Chronicle*, 30 December 1902, p. 5:3. Her children's column in the *Woman's Herald* had been written under the by-line, "Lizzie Bell." For an detailed exposition of Mrs. Stow and her activities, see: *The Life of Marietta Stow, Cooperator* ([Pacific Grove, California]: Pt. Pinos Editions, 1980, 244 pp.). Despite its vast amount of information on California women, caution in its use is advised because there is so much undocumented and inaccurate information as well as information that appears nowhere else. The book's original title was *Woman's Republic*. It was produced via photo-copying from a publisher's paste-up with the new title on the cover.

CHAPTER 9

1. *Carrier Dove*, 7 December 1889, p. 774:1.

2 . Robert D. Armstrong, *Nevada Printing History* (Reno: University of Nevada Press, 1981), p. 279.

3. San Francisco *Morning Call*, 31 July 1883, p. 4:1.

4. *Third Biennial Report of the Bureau of Labor Statistics of the State of California for the Years 1887-1888. John J. Tobin, Commissioner* (Sacramento: State Office. J. D. Young, Supt. State Printing, 1888), p. 14.

5. Ibid., p. 24.

6. Ibid., p. 77.

7. Ibid., p. 83.

8. *Fourth Annual Report of the Commissioner of Labor, 1888: Working Women in Large Cities* (Washington: Government Printing Office, 1889), p. 75

9. Ibid., p. 348.

10. *Farley's Reference Directory of Booksellers, Stationers and Printers in the United States and Canada* (Philadelphia: A. C. Farley & Co., 1885), p. 182.

11. [John J. Tobin], *Investigation by the Commissioner of the State Bureau of Labor Statistics into the Condition of Printing Establishments of San Francisco and Oakland, February and March, 1888* (Sacramento: State [Printing] Office, 1888). This is the source, passim, for data that follow.

12. Ibid., pp. 5–6.

13. Ibid., p. 6.

14. Fred H. Hackett, ed., *The Industries of San Francisco* (San Francisco: Payot, Upham Co., 1884), p. 121.

15. BLS *Investigation*, p. 15.

16. Ibid. San Francisco testimony, passim, pp. 8–13; Oakland, pp. 16–23.

17. *San Francisco Daily Examiner*, 25 February 1888, p. 4:6. This recalls Christopher Plantin's concern for his daughter in the proofing of his work (Chapter 1).

18. BLS *Investigation*, p. 15.

19. Ibid., p. 25.

20. KCBS-Radio (San Francisco), 27 May 1978. San Francisco newspapers apparently did not mention this suit.

21. *San Francisco Chronicle*, 15 June 1993, p. C1:1.

22. *Inland Printer*, April 1884, p. 10:2.

23. See Thomas Dublin, *Women at Work* (New York: Columbia University Press, 1979).

24. California Historical Society Library, San Francisco.

25. Elizabeth Faulkner Baker, *Displacement of Men by Machines* (New York: Columbia University Press, 1933), p. 5.

26. George E. Barnett, "The Introduction of the Linotype," *The Yale Review*, November 1904, pp. 251–273. Prof. Barnett 's monograph, "The Printers" (*American Economic Association Quarterly*, October 1909, 387 pp.) is a superb study of printing practice in the 19th Century, especially the role of trade unionism.

27. David I. Selvin, "History of the San Francisco Typographical Union" (M.A. diss., University of California, Berkeley, 1936), p. 52-A.

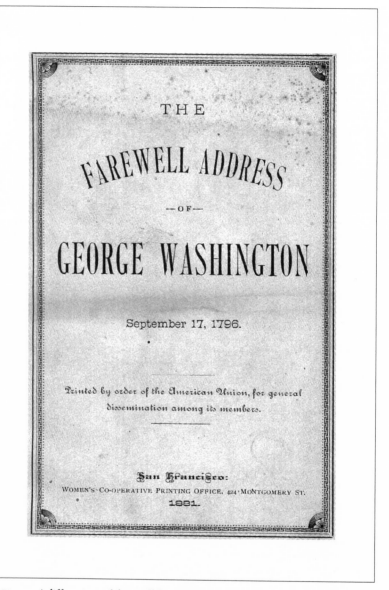

THE

FAREWELL ADDRESS

—OF—

GEORGE WASHINGTON

September 17, 1796.

Printed by order of the American Union, for general dissemination among its members.

San Francisco:

WOMEN'S CO-OPERATIVE PRINTING OFFICE, 424 MONTGOMERY ST.

1881.

Fig. 41:*A different use of the small format often used for constitutions and by-laws, with neat borders and several styles of type on the cover.* Author's Collection.

APPENDIX A

Reliability of Sources: Both Directories and the U. S. Census can be inconsistent and unreliable. Every effort has been made to verify entries of the most important women from whatever sources might be available. The fact that women in this era do not appear in Great Registers is a handicap as is the lack of readily available marriage records and other vital statistics.

Style used in all entries: The city of residence is abbreviated and changed in succeeding entries only if the city changed. Occupations are given *exactly* as found in the Sources and the city is noted as required. Data are linked to Sources by numerals within ornamented brackets, e.g., (0). Directory citations are Langley/San Francisco/Oakland unless otherwise indicated. Sequential, undesignated listings imply the same Directory and/or city and/or date as prior entry.

Scope of listing: The occupations noted, aside from printer, compositor, etc., are flexible so that women working in many printing-related tasks, such as lithographer, "works in type foundry" and "collector" (for woman-owned firms) are included. Binding is excluded because it was basically a separate trade by 1857, the beginning date of this study, and none of the woman-controlled establishments maintained a bindery.

Abbreviations: Investigation by the Commissioner of Labor Statistics into the condition of labor in printing establishments of Oakland and San Francisco. Feb. and Mar. 1888, John J. Tobin, commissioner (Sacramento, 1888), BLS Inv.; *Pacific Union Printer,* PUP; Alameda, (A); Berkeley, (B); Oakland, (O); San Francisco, (SF); San Jose, (SJ); Women's Co-operative Printing Union, WCPU. Typographical Union, TU; San Francisco Typographical Union #21, SF TU; Union records, UR; Labor Archives and Research Center, LA&RC.

Typographical Union notes: "Exempt" means a TU member is working at a job other than printing but retains Union status. A definition of "honorary member" by the SF TU has not been found. In Sacramento, the by-laws of TU #46, 1882, said: "Employing printers or others who stand fair with the craft or who have conferred benefits on the same, shall be entitled to honorary membership." Washoe (Nevada) TU #65, 1863, said that "Publishers, employers (being practical printers) or printers retired from the business, may be elected honorary members . . . ." It is possible that women so designated below had retired from the trade, as opposed to being on leave (i.e. exempt members) or they might have been widows of printers. None appear to have been master-printers. Fragmentary records of the SF TU for the period of this study are in the Labor Archives and Research Center, San Francisco State University, and in the *Pacific Union Printer,* in the Kemble Collections (the only complete file) at the California Historical Society Library.

Special Notes: (1) The Langley Directories were published from 1858–1895+; the Bishop Directories from 1875–1879. Researchers know that Directories are not always reliable. The most important persons in the listing have been checked from other sources where possible. As an example, Amanda M. Slocum's final married name remains unverified — Reid or Reed? — appearing in a Directory one way and in a book another. A problem of this kind is always discussed in the Note accompanying the entry. When Henry G. Langley took over *The San Francisco Directory for the Year 1858,* he noted (p. [11]), ". . . there was so much reluctance expressed by many persons to communicate necessary information. In this respect the compiler has met with so many cases where parties have refused their names, occupation and residence . . . ." Consequently, many women who worked in the printing trade are not traceable through Directories.

(2) PUP will be cited twice only if a job change had been made. Place of birth and age are from Census data unless otherwise noted. Because of inconsistencies in the 1870 and 1880 Census sheets, specific citations are omitted. The Ninth Census (1870) was especially useful in compiling the roster because "The age limitation of 15 years was. . . removed from the inquiry concerning the profession, occupation, or trade. . . ." As a result, several very young girls are listed.

(3) Whitton & Towne and Towne & Bacon woman employees have been included in the listing because examination of original records (1856–1875), now in the Special Collections Department of the Green Library, Stanford University, shows that many of them were often paid at the same rate as men who later became owners of their own printing firms in San Francisco. This alone would indicate the women were neither bindery labor nor in other lower-paying occupations which might occur in a printing-office. The Note that accompanies each of these entries gives some of the data from the records. Most importantly, Mrs. Sarah Maiers (A 138) is definitely the first woman of record (1857) to be employed by a San Francsco printing firm, some ten years after Samuel Brannan started the first printing operation in the town.

(4) For uniformity, the *year of publication* is cited for all Directories where given.

(5) A few women printers discovered working in places outside the immediate geographical areas of this study, such as Modesto and San Diego, are included as a matter of record so their identities will not be lost.

A ROSTER OF WOMEN IN PRINTING
Northern California, 1857–1890

1. Abraham, Nettie. 1305 1/2 Polk St. (SF). Collector with Mrs. Richmond [WCPU]. SOURCE: 1877 Bishop Directory.

Note: According to city Directories, newspapers and printers — among others — had "collectors." This person is the only one found for a woman's printing business and she immediately disappears from the Directory after the one entry. However, the manager and editor of *Golden Dawn* (monthly), Dora Darmoore, had a Miss Jennie L. Clarke as "business manager" (1877 Bishop Directory).

2. Allison, Jane A. Prospect Place (SF). Printer. Born: New York. Age: 32. SOURCE: 1880 Census.

Note: John Allison, 40, is listed as a printer living at the same address in the Census and John C., compositor, Bacon & Co., in the 1880 Directory.

3. Andler, Mary. [229] Tenth St. (SF). Compositor. Born: Pennsylvania. Age: 22. SOURCE: 1880 Census.

Note: Andlaur is entry for father, Jacques (James in Census), in the 1880 Directory.

4. Andornetti, Celeste. 826 Broadway (SF). Compositor, *Courier de San Francisco.* SOURCE: 1881 Directory.

5. Andrews, Mrs. Ada C. 941 Howard St. (SF). Compositor, San Francisco Newspaper Union. SOURCE: 1883 Directory

Note: The San Francisco Newspaper Union was a publishing business owned by Palmer & Rey.

6. Andrews, Miss G. Initiated into SF TU (1). Apprentice member, SF TU (2). SOURCE: (1) PUP, September 1889. (2) PUP, December 1889.

7. Apell, Miss Alma. 763 Seventeenth St. (O). Compositor, Pacific Press. SOURCE: 1889–90 McKenney Directory (O).

8. Auld, Frances. 2608 Leavenworth St. (SF) (1). Compositor, Mrs. L. G. Richmond & Son [WCPU] (1). Compositor, *Breeder & Sportsman* (2). Compositor, Bacon & Co. (2). Compositor, book & job offices. (3). SOURCES: (1) 1888 Directory. (2) BLS Inv. (3) PUP, October 1888.

Note: Testimony in BLS Inv., p. 10. Appears in SF TU "Membership Transition Records," 1888/1889 and "List of Members, Journeymen Only," 1 January 1910.

9. Ayers, Miss Edith. 1059 Castro St. (O). "With Pacific

Press." SOURCE: 1880 Bishop Directory (O).

10. Backess, Miss M. S. Honorary member, SF TU. SOURCE: PUP, October 1888.

Note: Charles E. Backess, compositor at the *Examiner,* residing at 322 Lombard St. (SF). SOURCES: 1886 and 1888 Directories. He was Chairman of Cemetery Committee of SF TU in PUP, October 1888. He appears as early as 8 September 1866 in the Receipt Book of Towne & Bacon (Vol. 28).

11. Backess, Mrs. Sarah E. Honorary member, SF TU. SOURCE: PUP, October 1888.

Note: Charles E. Backess, compositor at the *Examiner,* residing at 322 Lombard St. (SF). SOURCES: 1886 and 1888 Directories. He was Chairman of Cemetery Committee of SF TU in PUP, October 1888. He appears as early as 8 September 1866 in the Receipt Book of Towne & Bacon (Vol. 28).

12. Bailhache, Miss Mary. Composition, *Contra Costa Gazette.* SOURCE: 1884–85 McKenney Directory (O, A, B); listed under "Martinez, Contra Costa County."

Note: Same Directory shows a W. H. Brailhache [sic], notary public.

13. Barnes, Mrs. A. Apprentice member, SF TU. SOURCE: PUP, October 1888.

Note: A George E. Barnes is listed as an honorary member of the SF TU in the same issue.

14. Barrett, Kate S. Compositor, book & job offices (2). Initiated into SF TU (1). SOURCES: (1) PUP, March 1889. (2) PUP, July 1889.

15. Bayer, Mary F. Applicant, SF TU. SOURCE: PUP, May 1889.

16. Beagles, Miss. Lucy R. Two miles N of town, Napa City. Compositor, *Napa Register.* SOURCE: 1885 McKenney 8-County Directory.

17. Beatty, Miss Belle. 768 Twentieth St. (O). Compositor, Pacific Press. SOURCE: 1889 Directory (O).

Note: In the same Directory is Thomas Beattie, pressman, Pacific Press, and Mrs. Mary Beattie, widow, at the same address.

18. Bishop, Miss Emma. 153 16th St. (SF)(1). 574 E. 14th St. (O) (2). 618 E. 16th St. (O)(3). Compositor, Hollis & Carleton (1). Printer (2). Compositor, Pacific Press (3). Compositor, Pacific Press, formerly WCPU (4). SOURCES: (1) 1885 Directory. (2) 1886 Directory (O). (3) 1888 Directory (O). (4) BLS Inv. (O).

Note: Testimony in BLS Inv. (O), p. 21.

19. Blanchard, Miss Nonye. NW corner First St. & Fox Ave. (SJ). Compositor, *Daily Mercury.* SOURCE: 1884 Directory (SJ).

20. Blanchard, Miss Nora. 445 N First St. (SJ). Compositor, *Daily Mercury.* SOURCE: 1884 Directory (SJ).

21. Blanco, Simona. 516 Vallejo St. (SF). Compositor. SOURCE: 1882 Directory.

Note: Not in 1880 or 1881 Directories. Benicio Blanco, compositor, at same address in 1882 Directory.

22. Bland, Allie. 420 Locust St. (SJ). Printer. SOURCE: 1878 Directory (SJ).

Note: See Clara Bell Bland.

23. Bland, Clara Bell. Printer. SOURCE: 1878 Directory (SJ).

Note: Other San Jose Directories do not show her under Clara Bell. See Allie Bland.

24. Boland, Miss Ellen. 109 1/2 Shipley St. (SF). Compositor, Bacon & Co. SOURCES: 1881, 1882, 1883 Directories.

25. Bond, Mrs. Emma. Compositor, *Alta* (1). Card deposited from SJ TU #231 (1). Compositor, Hicks-Judd Co. (2). SOURCES: (1) PUP, December 1888. (2) PUP, August 1890.

Note: PUP, August 1889, lists her as Mrs. Emma Bond and in "Alta Chapel."

26. Boyd, Miss Emma. 708 11th St. (O) (1). Proofreader, Pacific Press (1)(2). SOURCES: 1886, 1887, 1888 Directories (O). (2) BLS Inv. (O).

Note: Testimony in BLS Inv. (O), p. 21.

27. Boyle, Mary. 409 Greenwich St. (SF) (1)(2). 313 Tehema St. (3). Compositor, Woman's Publishing Co. (1) Compositor, Cubery & Co. (2). Compositor (3). SOURCES: (1) 1874 Directory. (2) 1875 Directory. (3) 1877 Bishop Directory.

28. Brennan, Louise A. (1)(2). Miss Lou (3). 963 5th St. (O) (1)(2). Compositor (1)(2). Compositor, Pacific Press (3). SOURCES: (1) 1887–88 Directory (O). (2) 1888–89 Directory (O). (3) BLS Inv. (O).

Note: Testimony in BLS Inv. (O), p. 22.

29. Briggs, Mrs. Laura W. Compositor, *Woman's Herald of Industry.* SOURCE: *Woman's Herald of Industry,* October 1881, p. 2:1.

Note: Source gives details of her learning to set type. See Marietta Stow.

30. Brunner, Miss K. Compositor, book & job offices. SOURCE: PUP, December 1888.

Note: See Miss R. Brunner.

31. Brunner, Miss R. B. Card deposited from O TU #36, December 1888 (1). Compositor, book & job offices (2). Card withdrawn, SF TU (3). Card deposited, SF TU (4). Card withdrawn, SF TU (5). SOURCES: (1) UR, LA&RC. (2) PUP, February 1889. (3) PUP, June 1889. (4) PUP, July 1889. (5) PUP, December 1889.

Note: A Rudolph B. Brunner, 531 Sacramento St. (SF), compositor, *Examiner*, is listed in 1888–1890 Directories. See Miss K. Brunner.

32. Bryant, Emma. 3 1/2 Larkin St. (SF) (2). Type setter (1). Type setter, *California Farmer* (2). Born: California (1). Age: 26 (1). SOURCES: (1) 1880 Census. (2) 1881 Directory.

Note: See Louise Bryant.

33. Bryant, Miss Louise (1). E. Louise (2)(3). 1019 Bush St. (O) (1). Type writer and proof-reader, *Tribune* (1). Card deposited from O TU #36 (2). Compositor, book & job offices (3). SOURCES: (1) 1887 McKenney Directory (O). (2) PUP, October 1889. (3) PUP, November, December 1889.

Note: See Emma Bryant.

34. Buckner, Miss Belle M. 563 Second St. (SJ). Compositor, *Herald* office. SOURCE: 1878 Directory (SJ).

35. Burke, Miss Katie E. 718 Clementina St. (SF) (1)(2). Compositor (1). Compositor, Bacon & Co. (2) SOURCES: (1) 1882 Directory. (2) 1883 Directory.

36. Burke, Miss Kitty (1)(2)(3). Miss Kittie (4)(6). Miss Kate (5)(7). 528 Linden Ave. (1) 408 Fell St. (3)(4)(5)(6)(7). Compositor (1)(3). Compositor, C. W Nevin & Co. (2)(4)(5)(6)(7). SOURCES: (1) 1891 Directory. (2) *Morning Call*, 20 October 1892, p. 6:3. (3) 1894 Directory. (4) 1897, 1898 Directories. (5) 1899 Directory. (6) 1901 Directory (7) 1902 Directory.

Note: The column-long story about Miss Burke in the *Call* is headed, "A Very Fast Compositor. She is Famous for Speed and Accuracy. Employers Praise Her. Miss Kitty Burke Sets Fourteen Hundred Ems an Hour and Makes but Few Errors." From the 1891 Directory to the 1898 Directory, a Gilbert Burke is listed at the same addresses and is variously entered as a hostler, harness cleaner, stable man and watchman. In 1898 he evidently moved to 420 Fell Street. Kitty Burke disappears after the 1902 entry but Gilbert remains until 1914. He may have been her father or possibly a brother.

37. Burkett, Miss Clara B. 2nd & C Sts. (San Diego). Compositor, *Sun* office. SOURCE: 1886 San Diego Directory.

38. Byrne, Mrs. Laura. Applicant, SF TU. SOURCE: PUP, January. 1889.

39. Campbell, Florence E. SF TU applicant (1). Initiated, SF TU, 25 November 1888 (2) Compositor, book & job offices (3). SOURCES: (1) PUP, October 1888. (2) UR, LA&RC (3) PUP, January 1889.

Note: 1890 Directory: F. C. Campbell, compositor, 2 Olive Court (SF).

40. Carlisle, May A. 1361 Peralta St. (O). Compositor, Pacific Press. SOURCE: 1889 McKenney Directory (O).

Note: Herbert L. Carlisle, printer, Pacific Press, is also listed at the same address.

41. Chase, Caroline A. 512 1/2 Jones St. (SF). Compositor, WCPU. SOURCE: 1875 Directory.

42. Cleary, Mary. 543 Second St. (SF). Compositor, Bonnard & Daly. SOURCE: 1879 Directory.

43. Chaney, Mrs. Florence Wellman. Woman's Publishing Co. SOURCE: *Common Sense*, 13 February 1875, p. 464:2.

Note: When William and Amanda Slocum acquired the Woman's Publishing Co., they said, "The establishment is under the charge of Mrs. A. M. Slocum, the Business Manager of *Common Sense*, assisted by Mrs. Flora Wellman Chaney. Not only the management, but the book-keeping, soliciting and type-setting are all performed by women." There is no indication what specific duties fell to Mrs. Chaney, and there is no other record of her.

44. Cochell, Mattie E. (1)(2)(5). Miss Mamie (3) Miss M. E. (6)(10). Miss Mary E. (9)(10). 618 Fourth St. (SF) (1). 442 1/2 Natoma St. (2)(3)(4)(5). 1205 Golden Gate Ave. (6). 1913 Hyde St. (9)(10). Compositor, G. W. Henning (1). Compositor (2) Proofreader, Carlos White (3). Compositor, Carlos White (4). Proofreader, Pacific Newspaper Publishing Co. (5) Applicant, SF TU (7). Initiated into SF TU (8). Compositor, book & job offices (8). Compositor, *Post* ((9)(10). SOURCES: (1) 1875 Directory. (2) 1876 Directory. (3) 1879 Directory. (4) 1881 Directory. (5) 1882, 1883 Directories. (6) 1885 Directory. (5–7) PUP, May 1889. (8) PUP, June 1889. (9) 1890 Directory. (10) 1891 Directory.

Note: John T. Cochell, a compositor, is listed at Natoma St. addresses in Directories, passim, and appears as a Delegate to the Federated Trades in PUP, October, November 1888. The consistency of the two Cochells' street addresses confirms that the names listed for Mattie E. are probably variants for the same person. However, there is the possibility that there were three different persons residing at the same addresses.

45. Cochrane, Miss Mamie. 754 Mission St. (SF) (1). 772 Mission St. (2). 781 Mission St. (3). Compositor, Edward C. Hughes (1). Compositor (2)(3). SOURCES: (1) 1879 Directory. (2) 1881, 1882 Directories. (3) 1883 Directory.

46. Cole, Miss Kate. 927 Greenwich St. (SF). Compositor, Pacific Newspaper Publishing Co. SOURCE: 1877 Bishop Directory.

Note: Pacific Newspaper Publishing Co, was at 532 Clay St. which was for a brief period the address of *Golden Dawn,* Dora Darmoore's short-lived Spiritualist monthly. This building also had other printers as tenants. See Miss Nellie Cole.

47. Cole, Miss Nellie. 927 Greenwich St. (SF). Compositor, Pacific Newspaper Publishing Co. SOURCE: 1877 Bishop Directory.

Note: Pacific Newspaper Publishing Co, was at 532 Clay St. which was for a brief period the address of *Golden Dawn,* Dora Darmoore's short-lived Spiritualist monthly. This building also had other printers as tenants. See Miss Kate Cole.

48. Comstock, Mrs. Agnes M. 784 Folsom St. (SF) (1). 610 Front St. (SF) (3)(4). 411 Post St. (5)(7). Compositor, *Spirit of the Times* (1)(3). Compositor (2). Compositor, *Evening Post* (4)(5). Compositor, [Morning] *Post* (6). Compositor, *Post.* (7). Born: California (2). Age: 25 (2). SOURCES: (1) 1879 Bishop Directory. (2) 1880 Census. (3) 1883 Directory. (4) 1884 Directory. (5) 1886, 1887, 1888 Directories. (6) PUP, October 1888. (7) 1889 Directory.

Note: 1877 Bishop Directory: Mrs. A. M. Comstock at 401 Bryant St. (SF); 1878 Directory: no occupation, res. 784 Folsom St. Charles C. Comstock, compositor, *Spirit of the Times,* appears in Directory entries from 1883–1888 when name changes to Clarence B., working at *Bulletin.* Testimony in BLS Inv., p. 10. See Alice Comstock.

49. Comstock, Alice. Compositor, *Evening Post.* SOURCE: BLS Inv.

Note: See Mrs. Agnes Comstock.

50. Conroy, Miss Louise. 369 Fourth St. (O). Compositor. SOURCE: 1874 Directory (O).

51. Conway, Laura M. (1)(2). 114 Hayes St. (SF) (1)(2). Compositor (1). Compositor, R. G. Dun & Co. (2). SOURCES: 1888

Directory. 1889 Directory.

52. Coryell, Miss Lottie. 1843 Myrtle (O) {1}. 821 29th St. (O)
{2}. 665 20th St. (O) {3}. 1005 Webster St. (O) {4}. Compositor, *Oakland Times* {1}. Compositor, *Oakland Tribune* {2}{3}{4}{5}.
SOURCES: {1} 1884–85 Bishop Directory (O). {2} 1886 Directory (O). {3} 1887 Directory (O). {4} 1888 Directory (O). {5}
1888 INV (O).

Note: Testimony in BLS Inv. (O), p. 17.

53. Cosgro, Miss Louise. 124 Phelan Building (SF) {1}. Com-
positor, *Bulletin* {1}{3}. Card deposited from SJ TU #231 {2}.
Card issued, SF TU {2}. SOURCES: {1} 1890 Directory. {2}
PUP, October 1888. {3} PUP, November 1888.

Note: 1889 Directory lists her as "student" at same address.

54. Costello, Miss M. Nellie. Brown near Stewart, Napa City.
Compositor, *Napa Register.* SOURCE: 1885 McKenney 8-Coun-
ty Directory.

55. Dallas, Mrs. Amanda. Honorary member, SF TU.
SOURCES: UR, LA&RC, and PUP, October 1888.

Note: 1886 Directory: Widow, residence 18 Octavia St.

56. Dalt, Mary F. 2233 Howard St. (SF) {1}. 101 Precita Ave.
{2}. Compositor, A. H. Hollis {1}. Compositor {2}. SOURCES:
{1} 1888 Directory. {2} 1889 Directory.

57. Daly, Miss Mary F. Apprentice member, SF TU {1}.
Printer, *Chronicle* {2}{3}. SOURCES: {1} PUP, December 1889.
{2} PUP, August 1890. {3} 1891 Directory.

58. Delmer, Mrs. Louise R.; Miss Lizzie R. [Work? 1866],
Towne & Bacon. SOURCE: Pay Roll Book and Receipt Book,
Towne & Bacon, 1875–1873.

Note: The Towne & Bacon Pay Roll Book figures for each employee are entered under the following headings: "Regular Time, Rate Per Day, Extra
Time, Rate Per Week, Amount Due," followed by a space where each
signed for payment. There is no indication of typesetting being paid by
piecework, the established custom of the time. In general, women were averaging above the median for any given week's payroll, an indication they
were skilled labor, which probably meant typesetting. There is a confusion
in names in two entries that recur in both the Pay Roll Books and Receipt
Books. Two people are listed with Florence as a given name: Florence
O'Neil and Florence Yslas. Neither were women. Directories of the period
list them as pressmen who worked at various firms in San Francisco. Florence is unfamiliar as a male name, whereas Shirley, Marion and the like are
today better known in this context.

59. Denis, Miss Mary. 1413 Sacramento St. (SF). Printer, *Alta* job office. SOURCE: 1883 Directory.

60. Desrosier, Miss Ada. Paraguay St., near Railroad Ave. (SF). Compositor, Bacon & Co. SOURCE: 1883 Directory.

61. Devine, Mrs. K. Exempt member, SF TU. SOURCES: PUP, Ocober 1888 and UR, LA&RC.

Note: 1889 Directory: K. J. Devine, compositor, *Chronicle,* res. 551 1/2 Minna St.; also 1890. The 1888, 1889, 1890 Directories show a Robert J. Devine, compositor, *Chronicle,* living at 551 1/2 Minna St. 1886, 1887 Directories show him working at *Alta.*

62. Dewitt, Miss Lizzie. 14 Clinton St. (SF). Compositor. SOURCE: 1882 Directory.

63. Dexter, Miss Annie. East side East near South (SJ). Printer. SOURCE: 1878 Directory (SJ).

64. Dinan, Miss Johanna. 1 Scheerer St. (SF). Compositor. SOURCE: 1883 Directory.

65. Dodge, Miss Lydia Jennie. [Work? 1869], Towne & Bacon. SOURCE: Towne & Bacon Pay Roll Book, 1865–1873.

Note: She was paid for only two days @ $2.00 per day.

66. Donahue, Miss Agnes. 331 D St., Sacramento. Paper-ruler, Bernard Ruhl. SOURCE: 1889 Husted Sacramento Directory.

Note: Ruhl was a binder and blank-book manufacturer at 409 J St. Billheads, hotel registers, bookkeeping ledgers, journals, etc., all contained rules during this period that were created in a unique printing device, commonly called a ruling machine, which was sometimes L-shaped to facilitate the printing of rules at right angles. See Martin Heir, comp., *Twentieth Century Encyclopedia of Printing* (Chicago: Graphic Arts Publishing Co., 1930), pp. 365–370, for a full description and three illustrations. All the 19th Century San Francisco book and job shops run by women, or with woman employees, printed billheads and other forms which contained pen ruling.

67. Doyle, Miss Jennie. Applicant to Sacramento TU #46. SOURCE: Minutes of the Sacramento TU #46, v. 2:136, 26 August 1888.

Note: Minutes for this union are in The Bancroft Library. There is no further entry or action re Jennie Doyle recorded in the minutes.

68. Dowling, Miss Kate. 5 Page St. (SF){1}. Compositor, *Call*{1}{2}. SOURCES: {1} 1889, 1890 Directories. {2} PUP, October 1888.

Note: PUP, March 1889: served on SF TU funeral delegation.

69. Duckel, Alice. 916 Vallejo St. (SF) (1)(2). Compositor, Bacon & Co. (1). Compositor, *Occident* (3). Compositor, Francis & Valentine (2). Compositor, Bancroft's (3). Apprentice member of SF TU [to December 1889] (4). SOURCES: (1) 1884 Directory. (2) 1885, 1886, 1887, 1888 Directories. (3) BLS Inv. (4) PUP, October 1888.

Note: 1884 Directory: John, pressman, at same address. Both appear in Directories with same entries until 1888 when John is also listed as working at Francis & Valentine. Testimony in BLS Inv., p. 10.

70. Duenwald, Minna. Apprentice to printer. Born: New York. Age: 15. SOURCE: 1870 Census.

Note: A brother, Frank, 13, listed also as apprentice to printer.

71. Dugan, Adelia E. 317 Stockton St. (SF) (2). Compositor (1). Compositor, *New Age* (2). Exempt member, SF TU (3). Born: Iowa (1). Age: 21 (1). SOURCES: (1) 1880 Census. (2) 1880 Directory. (3) UR, LA&RC, and PUP, October 1888.

72. Dwyer, Katie. (1). Miss Kate (2). Miss Kittie (3). 10 1/2 Howard Court (SF) (1). Compositor (1). Compositor, *Alta* (2) (3). SOURCES: (1) 1884, 1886 Directories. (2) 1888 Directory. (3) BLS Inv.

Note: Testimony in BLS Inv., p. 9.

73. Eastaboo, Marge. Printer. Born: Illinois. Age: 20. SOURCE: 1870 Census (O).

Note: Lives with a sister ("at home"), 17, and other lodgers at the residence of James Siddon and wife.

74. Edgar, Maggie. NE corner Green & Montgomery Sts. (SF) (1). Compositor, Woman's Publishing Co (1). Printer (2). Born: California (2). Age: 22 (2). SOURCES: (1) 1875 Directory. (2) 1880 Arizona Census.

Note: A M. J. Edgar appears in the 1871 Directory as a compositor at the *Call*, residing at 17 1/2 Clinton St. (SF).

75. Eldridge, Clara. San Jose (1). Fourth St. (SF) (3). Type setter (2). Typesetter, A. M. Slocum, 612 Clay St. (3). Born: California (1)(2). Age: 7 (1), 17 (2). SOURCES: (1) 1870 Census. (2) 1880 Census. (3) 1880 Directory.

Note: Obituary, *San Jose Pioneer* (20 November 1880, p. 3:1), says she was compositor in printing office, "daughter of William N. Slocum, formerly of this city." See also Clara Slocum and Amanda Slocum.

76. Ellis, Miss Grace. 515 Sixth St. (SF) (1). Compositor,

John H. Knaiston (1). Compositor, Bacon & Co. (2). SOURCES: (1) 1887 Directory. (2) BLS Inv..

Note: Daily Evening Bulletin (25 February 1888, p. 4:2), reported that "Miss Grace Ellis and others gave testimony" at the BLS Inv. (p. 9).

77. Ellis, Mrs. W. W. Compositor; put on active list, Sacramento TU #46. SOURCE: Minutes of the Sacramento TU #46, v. 2:146, 25 November 1888.

Note: The minutes of this Union are in The Bancroft Library. Her husband was also activated at the same time, "said persons having gone to work at the case again."

78. Evans, Florence F. Member in good standing, SF TU, August 1888. SOURCE: UR, LA&RC.

Note: Name appears in "Membership Transition Records," 1888/1889. There is no female Evans in any PUP roster of the same period and none in city Directories.

79. Fairfield, Nellie E. 414 Beale St. (SF) (1). 317 Harrison St. (2). Compositor, Bacon & Co. (1). Compositor (2). Applicant to SF TU (3). Initiated into SF TU (4). Compositor, book & job offices (4). Compositor, *Call* (5). Compositor, Filmer-Rollins Electrotype Co (6). SOURCES: (1) 1883 Directory. (2) 1886, 1887 Directories. (3) PUP, June 1889. (4) PUP, July, 1889. (5) PUP, November, 1889. (6) PUP, November 1890.

Note: PUP, November 1889, reports her on SF TU's Membership Committee with Etta Mower (q.v.). They were the first women "officeholders" in SF TU.

80. Farwell, Harriet L. Market St. (SF). Printer. Born: California. Age: 18. SOURCE: 1880 Census.

Note: See Miss Hattie Farwell.

81. Farwell, Miss Hattie. 32 Turk St. (SF). Collector, Taylor & Nevin [printers]. SOURCES: 1881, 1882 Directories.

Note: 1881 Directory: Leonard N. Farwell, compositor, Taylor & Nevin; residing at 32 Turk St. See Harriett L. Farwell.

82. Fennel, Mary. [Printer?]. SOURCE: Mrs. Lizzie Richmond Judd's Will [see note].

Note: "To Mary Fennel for faithfulness to duty while in the employ of The Hicks-Judd Co., $250." Mrs. Judd divorced in 1888 and her firm, the WPCU, had been incorporated into Hicks-Judd, an oldtime bookbinding and printing firm, in 1887. The name WCPU was retained until a disastrous fire in 1901 but Mrs. Judd had no further connection with the business except as a corporate executive and stockholder until her death in 1898.

83. Ferguson, Jennie. Folsom St. (SF) (1). 734 Folsom St. (2).

Compositor (1). Compositor, Bacon & Co. (2). Born: Missouri (1). Age: 26 (1). SOURCES: (1) 1880 Census. (2) 1881 Directory.

84. Fisher, Miss Sarah J. 35 Valpariso St. (SF) (1). 1306 Jackson St. (3). Printer's Apprentice (1)(2). Compositor, WCPU (3). Born: New York (2). Age: 16 (2). SOURCES: (1) 1868 Bishop Directory. (2) 1870 Census. (3) 1871, 1872, 1873 Directories.

Note: She lived at same address as William Barry, pressman at *Alta*, as given in (1). He may have worked also at WCPU (1874). See Fisher interview in Chapter 7.

85. Fitzpatrick, Miss Kitty. 361 Tehama St. (SF). Type setter, Painter & Co. SOURCES: 1882, 1883 Directories.

86. Fleming, Abigail. Proofreader. Born: Massachusetts. SOURCE: 1870 Census.

87. Fleming, Miss K. A. [Work? 1867–1872], Towne & Bacon. SOURCE: Pay Roll Book, Towne & Bacon, 1865–1873.

Note: She worked steadily during this period with only one week's vacation, while most of the other woman employees came and went. Her pay rose over time from $2.00, to $2.50 to $3.00 per day and on three occasions Jacob Bacon himself signed the pay roll for her. The last time he did so, 18 May 1872, Ms. Fleming had worked only one day that week, and she never appears again in this Pay Roll Book.

88. Foran, Mrs. Elizabeth. 907 Bush St. (SF). Compositor, *Alta*. SOURCES: 1888, 1889 Directories.

Note: A Mrs. L. V. Foran (q.v.) is listed as an exempt member, SF TU, in PUP, October 1888. 1890 Directory: Mrs. Elizabeth Foran, "Widow, res. 919 Jones St." NB: not listed in *Alta* Chapel in PUP for 1888 or 1889.

89. Foran, Mrs. L. V. Exempt member, SF TU. SOURCES: UR, LA&RC, and PUP, October 1888.

Note: See also Mrs. Elizabeth Foran.

90. Ford, Miss A. M. 54 L St. between 2nd & 3rd, Sacramento. Compositor. SOURCE: 1876 Uhlhorn Directory (Sacramento).

Note: A search of the 1876 Sacramento Directory produced this single name. Alice Woodeson (q.v.) was the first woman admitted to the Sacramento TU in 1888.

91. Ford, Miss Nellie. 1230 Pacific Ave. (SF). [Press] feeder, The Bancroft Company. SOURCE: 1889 Directory.

92. Ford, Nettie 1018 Twelfth St. (O) (1). 1316 Adeline St. (2). 533 Sixteenth St. (3). Compositor, Pacific Press (1)(2)(3).

SOURCES: (1) 1887 McKenney Directory (O). (2) 1888 Directory (O). (3) 1889 Husted Directory (O).

93. Forde, Mrs. M. L. (3). Ford, Miss M. L. (2). Ford, Miss May L. (1). Alameda (4). Compositor, Bacon & Co. (1). Compositor, *Post* (2)(3). Exempt member, SF TU (5). SOURCES: (1) *The Daily Examiner*, 25 February 1888, p. 4:6. (2) BLS Inv.. (3) PUP, October 1888. (4) 1888, 1889 Directories (SF). (5) PUP, August 1890.

Note: Mrs. Forde appears until at least December 1888 in PUP as working at the *Post*. A William C. Forde also appears in Directories as compositor at the *Post* during this period and living on Chapin St., Alameda. The confusion in the spelling of the last name comes from sources where both versions appear passim; Forde is correct. 1892–93 Husted Directory (O, A, B) shows William C. Forde as printer (SF) and living at 1818 Chapin St., Alameda. At same address is Hyacinthe Forde, printer (SF), not found elsewhere. Testimony in BLS Inv., p. 9.

94. Fulton, Josephine. NE corner Illinois & Sierra Sts. (SF). Compositor, *Occident*. SOURCE: 1875 Directory.

95. Gallagher, Miss C. Compositor, *Bulletin*. SOURCE: PUP, October 1888.

96. Gallagher, Miss Nellie. 315 Hunt St. (SF). Typesetter with California Type Foundry Co. SOURCE: 1872 Directory.

97. Gallagher, Ellen. Printer. Born: Kentucky. Age: 12. SOURCE: 1870 Census.

98. Gaynor, Miss Ann. 227 Perry St. (SF) (1). Compositor (1). Book & job offices (2). SOURCES: (1) 1887 Directory. (2) PUP, August 1890.

99. Gaynor, Mamie (1)(3)(4)(5). Miss Mary (2). 227 Perry St. (SF) (1)(2)(4)(5). Compositor (1)(2)(5). Book & job offices (3). Printer, SF TU (4). SOURCES: (1) 1886 Directory. (2) 1889 Directory. (3) PUP, August 1890. (4) 1890 Directory. (5) 1891 Directory.

100. Geddes, Miss Euphemia. 33 Moss St. (SF). Typesetter, Painter & Co. SOURCE: 1885 Directory.

101. Gegan, Angeline. Union St. (SF). Type setter. Born: Massachusetts. Age: 15. SOURCE: 1880 Census.

Note: See Lillie C. Gegan. The Gegans were sisters-in-law according to Census.

102. Gegan, Lillie C. Union St. (SF). "Works in printing office." Born: Massachusetts. Age: 17. SOURCE: 1880 Census.

Note: See Angeline Gegan. The Gegans were sisters-in-law according to Census.

103. Geohagan, Annie. Compositor, book & job offices (3). Applied to SF TU (1). Initiated into SF TU (2). SOURCES: (1) PUP, April 1889. (2) UR, LA&RC and PUP, May 1889. (3) PUP, June 1889.

104. Gilfillan, Mrs. S. 137 Third St. (SF). Compositor, Bacon & Co. SOURCE: 1883 Directory.

105. Gitt, Miss Lola. Compositor, book & job offices (1). Card deposited from Los Angeles TU #174 (2). SOURCES: (1) PUP, April. 1889. (2) PUP, May 1889.

106. Gorman, Miss Nellie (1)(2)(3). 2506 Post St. (SF) (1). 2508 Bush St. (2). Compositor, Carlos White Printing Co. (1). Compositor, Mrs. L. G. Richmond & Son (WCPU) (2) Compositor, Carrier Dove Printing & Publishing Co. (3). SOURCES: (1) 1886 Directory. (2) 1888 Directory. (3) 1890 Directory.

Note: The 1885 Directory shows a Thomas Gorman as compositor with A. L. Bancroft & Co., living at 525 Franklin St. There are other Gorman males listed as printers in SF Directories.

107. Gorman. Nellie (2)(3)(8)(9). Gorham [Gorman], Nellie (4). Gorman, Helen [Nellie] (5). Miss Nellie (10). Miss Nellie F. (1)(7). Nellie F. (6). E. Oakland (1). 477 Eleventh St. (O) (2). 477 E. Eleventh St. (3)(4)(5). 680 10th St. (6)(7)(8)(9). 1123 Myrtle St. (10). Compositor, Bacon & Co. (SF) (1). Compositor, Pacific Press (2)(3)(4)(5)(6)(7)(8)(9)(10). SOURCES: (1) 1888 Directory. (2) 1888 McKenney Directory (O). (3) 1889 Husted Directory (O). (4) 1890 Husted Directory. (5) 1891 Husted Directory. (6) 1892 Husted Directory. (7) 1893 Husted Directory. (8) 1894 Husted Directory. (9) 1896 Husted Directory. (10) 1897 Husted Directory.

Note: The Nellie Gormans enumerated here have been separated by their San Francisco and Oakland entries because there may have been two Nellie Gormans working simultaneously. The conflicting evidence centers on Nellie F., the "Oakland Gorman," who (according to her descendants) met and married her employer's son when she worked at *The Carrier Dove* printing-office. and was raising a family in SF during the years she is shown working and living in Oakland.

108. Grady, Miss Mary H. (4). Miss M. (1)(2). M. H. (3). 12 Moss St. (SF) (1)(3)(4). Compositor, *Call* (1)(2)(3)(4). SOURCES: (1) 1888 Directory. (2) PUP, October 1888. (3) 1889 Directory. (4) 1890 Directory.

109. Green, Catherine. 230 Hickory Ave. (SF) {1}. Compositor, *Evening Post* {1}. Works in printing office {2}. Age: 29 {2}. SOURCES: {1} 1881 Directory. {2} 1880 Census.
Note: See Mrs. K. Green.

110. Green, Miss E. C. 532 N. Third St. (SJ). Printer, *San Jose Daily Herald.* SOURCE: 1887 Ulhorn & McKenney Directory (SJ).

111. Green, Mrs. K. Compositor, book & job offices {2}. Suspended by SF TU {1}. SOURCES: {1} PUP, Ocober 1888. {2} PUP, December 1888.
Note: See Catherine Green.

112. Greene, Mrs. Kate. 215 Fourth St. (SF). Compositor, *Bulletin.* SOURCE: 1883 Directory.

113. Gregg, Louisa L. {1}. Louise {2}. 990 Willow St. (O). {1}{2}. Compositor {1}. Compositor, *Christian Independent* {2}. Compositor, book & job offices {4}. Card deposited from O TU #36 {3}. Withdrew from SF TU {5}. SOURCES: {1} 1887 McKenney Directory (O). {2} 1888 McKenney Directory (O). {3} PUP, May 1889. {4} PUP, June 1989. {5} PUP, August 1889.
Note: PUP reports "trade dull" {5}. Lived at same address as Mary A. Gregg.

114. Gregg, Mary A. {2}{3}{4}. Miss Mattie A. {1}{5}. 990 Willow St. (O) {1}. Compositor {1}. Compositor, book & job offices {3}. Compositor, *Christian Independent* {5}. Card deposited from O TU #36 {2}. Withdrew from SF TU {3}. SOURCES: {1} 1887 McKenney Directory (O). {2} PUP, May 1889. {3} PUP, June 1889. {4} PUP, August 1889. {5} 1889 Husted Directory (O).
Note: By 1891 it is Miss M. A. Gregg, printer, and she was working at the Tribune Publishing Co. and living at 1719 Eighth St. (O). Lived at same address as Louisa L. Gregg.

115. Gunterman, Alice. 1570 Seventh St. (O). Compositor {1}. Withdrew card from SF TU {2}. SOURCES: {1} 1889–90 McKenney Directory (O). {2} PUP, April 1889.
Note: "Membership Transition Records," 1888/1889, at LA&RC show her as a transfer from Seattle #202, that she was suspended in February 1889 and her SF TU card was issued in March 1889.

116. Harlow, Fayette M. {1}{2}{3}{5}{7}. Fay M. {4}. 37 Natoma St. (SF) {1}. 862 Howard {2}. 874 Howard {3}. 862 Howard {4}{5}. Printer {1}. Compositor, *Examiner* {2}{3}{6}. Apprentice member SF TU {4}. Compositor {5}. Compositor, *Examin-*

er (7). SOURCES: (1) 1886 Directory. (2) 1887 Directory. (3) 1888 Directory. (4) PUP, October 1888. (5) 1889 Directory. (6) PUP, April 1889. (7) 1890 Directory.

Note: 1886 Directory: Claude, printer, 379 Natoma St. 1887 Directory: Claude, compositor, Francis, Valentine. 1888 Directory: Claude, compositor, *Chronicle*, 874 Howard St. 1889 Directory: Claude, compositor, 862 Howard St. 1890 Directory: Claude, Plumber, 862 Howard St.; Joseph C., compositor, *Chronicle*, 862 Howard St.

117. Harty, Lizzie V. South side Turk, near Buchanan St. (SF) (1). 233 Third St. (SF) (2)(3). 420 Montgomery St. (SF) (3)(4). Compositor, WCPU (1). Compositor, *Saturday Evening Mercury* (2). Compositor, *Pioneer* (3)(4). Compositor, *Alta* (5). SOURCES: (1) 1868 Directory. (2) 1869 Directory. (3) 1871 Directory. (4) 1872 Directory. (5) 1873 Directory.

118. Hayden, Miss May. Compositor, book & job offices (2). Card deposited from Multnomah TU #58 (Portland, OR) (1). SOURCES: (1) PUP, January 1889. (2) PUP, February 1889.

119. Hazelbecker, Miss Clara. NE corner Park Ave. & Orchard St. (SJ). Printer. SOURCE: 1878 Directory (SJ).

120. Healey, Miss Fannie. Compositor, Pacific Press ["... worked at other offices"]. SOURCE: BLS Inv. (O).

Note: Testimony in BLS Inv. (O), p. 22.

121. Henry, Miss Minnie. 26 East St. (SJ). Compositor. SOURCE: 1887 Ulhorn & McKenney Directory (SJ).

122. Hickey, Miss Maggie. Compositor, book & job offices (2)(4). Compositor, *Chronicle* (3). Initiated into SF TU (1). Suspended from SF TU, 1890. (5). SOURCES: (1) PUP, October 1888. (2) PUP, November 1888. (3) PUP, December 1888. (4) PUP, March 1889. (5) PUP, August 1890.

123. Higbee, [Mrs.] L. P. (1). Mrs. L. (2). 31 Eddy St. (SF) (2). Keeping House (1). Printer (2) Born: New Hampshire (1). Age: 31 (1). SOURCES: (1) 1870 Census. (2) 1872 Directory.

Note: Higbee was the married name of Lisle Lester (q.v.) when she returned to SF but there is no record of her working as a printer. She is listed in the Census under her husband's entry; hence, "L. P." Lyman Higbee was an attorney and appeared in the SF Directory under "Attorneys" in 1871 only. The Census gives his age as 46 and New York as his place of birth. All their parents were native-born. (Reference: 9th Census, San Francisco, Reel 83, Precinct 3, Ward 10, p. 40.)

124. Hoffman, Elizabeth. Lithographer's apprentice. Born:

Germany. Age: 19. SOURCE: 1870 Census.

Note: A brother, 17, is listed in the Census as a printer's apprentice.

125. Horton, Miss Rachel. High St. near Jefferson St. (A) (1)(3). Alameda (2). Compositor, Pacific Press (1). Compositor, *Enquirer* (2)(3)(4). SOURCES: (1) 1883 McKenney Directory (O). (2) 1887 Directory (O). (3) 1888, 1889 McKenney Directories (O). (4) BLS Inv. (O).

Note: Testimony in BLS Inv. (O). p. 22.

126. Humphrey, Amy C. Compositor, *Morning Call.* SOURCE: 1871 Directory.

127. Humphreys, Anne. Compositor. Born: New Hampshire. Age: 21. SOURCE: 1870 Census.

Note: Census shows that she lived at the same boarding house as Emma Hutchinson.

128. Hutchinson, Emma. 517 Clay St. (SF) (1)(3). Compositor, WCPU (1)(2). Compositor, *Morning Call* (3). Born: Pennsylvania (2). Age: 25 (2). SOURCES: (1) 1868 Directory. (2) 1870 Cenus. (3) 1871 Directory.

Note: Census shows that she lived at the same boarding house as Anne Humphreys.

129. Hutchinson, Miss M. F. 233 Third St. (SF). Compositor, WCPU. SOURCE: 1869 Directory.

Note: See Mrs. Margaret Hutchinson.

130. Hutchinson, Mrs. Margaret ("widow"). 743 Vallejo St. (SF). Typesetter. SOURCE: 1874 Directory.

Note: See Miss M. F. Hutchinson.

131. Hutchinson, Miss S. A. Card deposited from Los Angeles TU #174. Transferred to exempt list, SF TU. SOURCE: PUP, December 1889.

132. Hutchinson, Miss Sarah A. 137 Third St. (SF). Compositor. SOURCE: 1883 Directory.

133. Ireland, Miss Jennie. 708 11th St. (O). Type-setter, Pacific Press (1)(2). SOURCES: (1) 1886, 1887, 1888 Directories (O). (2) BLS Inv. (O).

Note: Directories show John J. Ireland, booking clerk, Pacific Press, at same address as well as a Mrs. May Ireland. Testimony in BLS Inv. (O), p. 20.

134. Irvin, Miss Cecelia. [Work? 1867], Towne & Bacon. SOURCE: Pay Roll Book, Towne & Bacon, 1865–1873.

Note: This woman worked briefly at $1.00 per day. Was she was an apprentice typesetter or did she perform less skilled chores in the printing-office? are questions unanswered by the data.

135. Irving, Miss Maggie B. (1)(2)(4). Miss M. (3). 1047 1/2 Harrison St. (SF)(1). 762 Bryant St. (4). Compositor, *Bulletin* (1)(3)(4). Compositor (2). SOURCES: (1) 1887 Directory. (2) 1888 Directory. (3) PUP, October 1888. (4) 1889, 1890 Directories.

Note: The Daily Examiner (25 February 1888, p. 4:6) reported Miss Maggie Irving as giving testimony at the BLS Inv. (p. 9). She was incorrectly designated as Maggie Irwin in the BLS Inv.

136. Johnson, Elizabeth J. Howard St. (SF). Compositor. Born: California. Age: 18. SOURCE: 1880 Census.

137. Johnson, Miss Nellie. N side San Antonio between Second & Third Sts. (SJ). Printer, *Herald* Office. SOURCE: 1878 Directory (SJ).

138. Johnston, Miss E. C. 13 S. Fifth St. (SJ). Printer. SOURCE: 1887 Ulhorn & McKenney Directory (SJ).

139. Jollymour, Miss Minnie. Park ave. near the bay (A). Printer, SF. SOURCE: 1889 McKenney Directory (O, A, B).

140. Jorgensen, Rose. 1303 Twenty-fourth Ave. (O). Compositor. SOURCE: 1887 Husted Directory (O).

141. Kanary, Hannah. [Work? 1858], Towne & Bacon. SOURCE: Receipt Book, Vol. 28, Towne & Bacon Records, Stanford University.

Note: She received $15.00 a week for three of the four weeks for which she was paid in October. Because of the hours and scale it is probable that she was a typesetter.

142. Kennedy, Miss Josephine. 32 Moss St. (SF)(1)(2). Compositor, Bacon & Co. (1). Compositor (2). Initiated, SF TU, October. 1890 (3). SOURCES: (1) 1889, 1890 Directories. (2) 1891 Directory. (3) PUP. October 1890.

143. Kerr, Miss Clara. Compositor, book & job offices (1). Printer, Filmer-Rollins Electrotype Co. (2). SOURCES: (1) PUP, December 1889. (2) 1891 Directory.

144. King, Mary. 1307 Montgomery St. (SF)(1). Typesetter, Painter & Co. (1). Typesetter (2). Born: Massachusetts (2). Age: 18 (2). SOURCES: (1) 1880 Directory. (2) 1880 Census.

145. Lafontaine, Mrs. A. J. 625 Merchant St. (SF). Printer.

Source: *Farley's Reference Directory of Booksellers, Stationers and Printers in the United States and Canada* (Philadelphia: A. C. Farley & Co., 1885), p. 18:2.

Note: A. J. Lafontaine had been a job printer in San Francisco since 1858. This is the only reference found indicating his widow continued the business under her own name, a first for San Francisco. *Farley's* estimated the established capital of her firm as $500. (compared to "$10–20,000" for Mrs. L. G. Richmond [WCPU]), and her credit was listed as "moderate" (Mrs. Richmond's, "good").

146. Leonard, Mrs. J. T. Membership in SF TU renewed. SOURCE: PUP, October 1889.

147. Lester, Lisle [pseud.] (1)(2)(3). Sophia Emeline Walker [given name] (2). Mrs. L[yman] P. Higbee [second marriage] (2)(4). Lester, Miss Lisle (3). 327 Bush St. (SF) (3). 31 Eddy St. (4). Proprietress, *Pacific Monthly* (1). Printer (4). Editor, *Pacific Monthly*; employed women printers (5). A founder of the Female Typographical Union (6). Born: New Hampshire (2). Age: 23 (1860) (2). SOURCES: (1) 1864 Directory. (2) *NEWMONTH*, October 1984, pp. 13–17. (3) 1865 Directory. (4) 1872 Directory. (5) *The History of Woman Suffrage*, Stanton et al., v. 3:762. (6) *Pacific Monthly*, May 1864, p. 622.

Note: According to Stanton et al., Lisle Lester was the first to hire women printers (i.e. typesetters) in San Francisco. After leaving the city, she remarried and returned as Mrs. Lyman P. Higbee but listed herself in the 1869 and 1870 directories as "Mrs. Lisle Higbee, furnished rooms, 313 Jessie St." There is no entry for her in 1871 but one for "Liman [sic] P. Higbee, attorney and counselor at law, 313 Jessie St." He is also listed under "Attorneys" in the same Directory but disappears thereafter. Although Mrs. Higbee is listed as a printer in the 1872 Directory, no record of her employment in the trade at that time has been found. Her entry under Miss Lisle Lester in the 1865 Directory (p. 52) was a belated one for she appears under "Additional Names . . ." and the reason may well have been a recent incapacitation. See also Mrs. L. P. Higbee.

148. Levy, Miss Ida. 30 Park Ave. (SF). Compositor, Bacon & Co. SOURCES: 1885, 1886, 1887 Directories.

149. Levy, Miss Lillie. 30 Moss Ave. (SF) (1). Compositor, Bacon & Co. (1). Compositor, Filmer-Rollins Electrotype Co. (2). SOURCES: (1) 1885 Directory. (2) PUP, November 1890.

150. Little, Jennie L. 1022 1/2 Larkin St. (SF) (2). 212 Powell St. (4) Printer (1). Compositor, *Evening Bulletin* (2). Compositor, *Chronicle* (3)(4). Born: New York (1). Age: 23 (1).

SOURCES: (1) 1880 Census. (2) 1881 Directory. (3) PUP. October 1888, August 1889. (4) 1889 Directory.

151. Lomler, Inez. Bernard St. (SF) ("boarder"). Printer. Born: Kentucky. Age: 21. SOURCE: 1870 Census.

152. Loughborough, Maria. 708 Eleventh St. (O). Compositor, Pacific Press. SOURCE: 1887 McKenney Directory (O).

Note: At same address were: Miss May, folder at Pacific Press; Delmer, a printer at Pacific Press; and J. N. Loughborough, clergyman.

153. Loveland, Dora I. 1507 Folsom St. (SF) (1). Tehama St. (2). Compositor, Dore & Co. (1). Printer (2). Age: 17 (2). SOURCES: (1) 1880, 1882 Directories. (2) 1880 Census.

154. Ludwig, Miss Rosa. 806 E. Fourteenth St. (O). Printer, W. T. Bailey. SOURCE: 1887 McKenney Directory (O).

Note: 1874 Directory shows Wm. T. Bailey, job printer, *Oakland Home Journal* [and *Alameda City Advertiser*], Wm. Halley editor and proprietor. This might be an example of a job shop being run separately or specifically to print a newspaper or journal. Other Directories show Bailey only as "printer."

155. Maiers, Mrs. Sarah. [Work? 1857], Whitten & Towne. SOURCE: Whitten & Towne Receipt Book, Vol. 26, Towne & Bacon Records, Stanford University.

Note: Her pay records, which she signed, covered the weeks ending 6 June to 18 July 1857. The first entry shows she received $22.50 for nine days' pay. Thereafter, she never exceeded $15.00 for a 60-hr. week, at $2.50 per day, which was comparable to the rate many of the men were paid, and more than some. Mrs. Sarah Maiers was most probably the first woman typesetter of record in San Francisco.

156. Marietta, Constance. Apprenticed to printer. Born: California. Age: 11. SOURCE: 1870 Census.

157. Marshall, Anna (2). Annie (1). 1718 Jones St. (SF) (1). Jones St. (2). Typesetter, Painter & Co. (1). Printer (2). Born: Canada (2). Age: 17 (2). SOURCES: (1) 1880 Directory. (2) 1880 Census.

158. Mason, Emma L. (1). Miss E. L. (3). 415 Powell St. (SF) (1). Compositor, *Bulletin* (4). Compositor, Carlos White's (2). Compositor, Parsons' (2). Compositor, *Bulletin* Office (3). Compositor, *Bulletin* (2) (3). SOURCES: (1) 1886 Directory. (2) BLS Inv. (3) PUP, October 1888. (4) 1889 Directory.

Note: Testimony in BLS Inv., p. 10.

159. Mavity, Katie. 714 Eleventh St. (O). Compositor, Pacific

Press. SOURCE: 1887 McKenney Directory (O).

Note: 714 Eleventh St. was the address of the Pacific Press Boarding House, according to the same Directory.

160. McDermott, Mary (1). Mamie (2) (3). 1014 Filbert St. (SF) (1)(3). Compositor, Bacon & Co. (1). Francis Valentine & Co. (2)(3). SOURCES: (1) 1888 Directory. (2) PUP, August 1890. (3) 1891 Directory.

161. McDonald, Miss Minnie. Compositor, book & job offices (1). Compositor, *Post* (2). Card deposited from Sacramento TU #46 (1). Exempt member, SF TU (3). SOURCES: (1) PUP, November 1888. (2) PUP August 1889. (3) PUP, August 1890.

Note: Sacramento TU #46 minutes, v. 2:139 (30 September 1888) say "card received: Miss Minnie McDonald." These minutes are in The Bancroft Library. Minnie McDonald had also deposited her card from Los Angeles TU #174 (PUP, 12 July 1889, 4:1).

162. McFarland, Minnie (1). [Miss Mary (3)]. 2121 Jones St. (SF) (1). Compositor (1). Compositor, book & job offices (3). Printer, Filmer-Rollins Electrotype Co. (4). Applied to SF TU (2). Initiated into SF TU (3). SOURCES: (1) 1887 Directory. (2) PUP, May 1889. [(3) PUP, June 1889.] (4) 1891 Directory.

Note: Erroneously listed as Mary in PUP but is in fact as (1) above.

163. McGee, Miss Mary. 211 Minna St. (SF) (2). First president of Female TU (1). Compositor, *Christian Advocate* (2). SOURCES: (1) *Pacific Monthly*, May 1864, p. 622. (2) 1865 Directory.

Note: Writing in *Pacific Monthly*, Lisle Lester complimented "our lady compositor" whom she called "McKee." This was probably a typographical error for Mary McGee as no McKee is in any Directory — *Pacific Monthly* was uncommonly full of typographical errors. McGee is not in 1863, 1864 or 1866 Directories as a compositor. The Female TU was organized 25 May 1864 at the *Pacific Monthly* with Mary McGee as president. See also Mary E. Parker.

164. McGonagle, Rosanna. NW Corner Clay and Drum Sts. (SF). Works in Type Foundry. Born: Massachusetts. Age: 16. SOURCE: 1870 Census.

Note: 1869 Directory lists mother, Catherine, 40, as McGonigle.

165. McGonagle, Teresia. NW corner Clay and Drum Sts. (SF). Works in Type Foundry. Born: Massachusetts. Age: 18. SOURCE: 1870 Census.

Note: See note to Rosanna McGonagle.

166. McGowan, Miss M. A. Typesetter, *Evening Post* (1). Typesetter, *Morning Post* (2). SOURCES: (1) BLS Inv. (SF). (2) PUP, October 1888.

Note: Testimony in BLS Inv., p. 9. 1880 Census showed Margaret, 12, as a sister of Mary McGowan (q.v).

167. McGowan, Miss Mary. 22 1/2 Lewis Place (O) (2). Typesetter, *Evening Post* (1). Typsetter, *Morning Post* (2). SOURCES: (1) BLS Inv. PUP. October 1888.

168. McGowan, Miss Mary. 22 1/2 Lewis Place (O) (2). Typestter (1). Compositor, *Post* (2)(3). Born: California (1). Age: 17 (1). SOURCES: (1) 1880 Census (O). (2) 1887, 1888 Directories (O). (3) 1889 Directory (O).

Note: See Miss M. A. McGowan.

169. McKay, Mary C. 1103 Powell St. (SF) (1). Mason St. (2). Compositor, B. H. Daly (1) Compositor (2). Born: California (2). Age: 18 (2). SOURCES: (1) 1880 Directory. (2) 1880 Census.

Note: Lived with mother, sister and aunt on Mason St., according to Census.

170. McKean, Miss Anna (1). Miss Annie (2). 518 Eddy St. (SF) (1)(2). Compositor, *Occident* (1)(2). SOURCES: (1) 1872 Directory. (2) 1873 Directory.

Note: This may be the same person as Anna M. McKean of San Jose (q.v.) but no proof has been found.

171. McKean, Anna M. Sunol near South St. (SJ). Compositor, *California Agriculturist.* SOURCE: 1876 Directory (SJ).

Note: This may be the same person as Miss Anna McKean of San Francisco (q.v.) but no proof has been found.

172. McLane, Mary. 754 Folsom St. (SF) (1) (2). Compositor, Woman's Publishing Co. (1). Compositor, Cubery & Co. (2). SOURCES: (1) 1875 Directory. (2) 1876, 1877 Directories.

173. McLean, Alice. 1429 Mission St. (SF) (1). 15 1/2 Grand Ave. (2). Compositor, WCPU (1). Compositor, *Occident* (2). Compositor, *Call* (3) (4). SOURCES: (1) 1872, 1873 Directories. (2) 1874, 1875, 1876, 1877 Directories. (3) 1888 Directory. (4) PUP, October 1888.

174. McLean, Miss Myra (1). 2724 Seventeenth St. (SF) (1)(2). Compositor, San Francisco Newspaper Union (1). Compositor, Palmer & Rey (2). SOURCES: (1) 1883 Directory. (2) 1886

Directory.

Note: Palmer & Rey were the owners of the San Francisco Newspaper Union, a publishing business, as well as typefounders and printing equipment dealers.

175. Mills, Sadie. [724 Folsom St. (SF).] Compositor. Born: Oregon. Age: 16. SOURCE: 1880 Census.

176. Moody, Miss Maggie M. 320 E. Twenty-second St. (O) Compositor, Carruth & Carruth. SOURCE: 1889 McKenney Directory (O, A, B).

Note: Miss Maggie Moody, "res cor E Twenty-second and Seventh Ave," appears in 1887 McKenney Directory (O) without occupation; "res SE cor Twenty-second and Seventh Ave." in 1888 Directory, (name as "M. M." only), no occupation given.

177. Moore, Marianne. [Work? 1858], Whitten & Towne. SOURCE: Receipt Book, Whitten & Towne, Vol. 26, Towne & Bacon Records, Stanford University.

Note: She was shown as receiving pay for weeks ending 11 and 23 July 1858, $8.50 and $5.00 respectively but she never worked more than a few hours.

178. Moran, Miss Christine M. (1). Miss Chrissie (2)(3). 1113 Montgomery St. (SF) (1). Compositor (1). Compositor, book & job offices (2). Compositor, *Post* (3). Initiated into SF TU (2). SOURCES: (1) 1888 Directory. (2) PUP, June 1989. (3) PUP, November 1889.

Note: Chrissie Moran of *Post* was elected to Membership Committee of SF TU, vice Etta Mower, in January 1890 (PUP).

179. Moran, Miss L. Exempt member, SF TU. SOURCE: PUP, November 1888.

Note: 1880 Census has a Lottie Moran, 7, at same address as Susie Moran (q.v.).

180. Moran, Miss Susan E. (1). Miss Susie (3)(4)(5)(9). Susan (5)(7). 112 Geary St. (SF) (1). 550 Natoma St. (3)(4)(5)(6) (7). 1220 Leavenworth St. (9). Compositor, WCPU (1)(2)(3). Compositor, *Morning Call* (4)(6). Compositor, *Call* (5)(7)(9). Exempt member, SF TU (10). Born: Massachusetts (8). Age: 23 (8). SOURCES: (1) 1869, 1870, 1871 Directories. (2) *Woman's Pacific Coast Journal*, August 1870, p. 58:3. (3) 1872 Directory. (4) 1873, 1874 Directories. (5) 1875 Bishop Directory. (6) 1876 Directory. (7) 1878 Directory. (8) 1880 Census. (9) 1886 Directory. (10) PUP, October 1883.

Note: The 1887 Directory (O) shows Misses Mary and Susie Moran, milliners

("The Misses Moran"), at 1452 Seventh St. At the time of the 1880 Census, Mary (Marie in Census) was a year older than Susie Moran and listed as a milliner. Tracing her in Directories shows that she had worked as early as 1870 for others as a milliner in San Francisco and finally at 1452 Seventh St., Oakland. In the 1887 Directory (O) Susie Moran was residing in San Francisco. The 1888 McKenney (O) Directory has "Miss Susie Moran, milliner, 1452 Seventh St." Apparently their mother, "Lizzie" in Directory (Elizabeth in Census), joined the sisters as a milliner in Oakland as well and 1452 Seventh St. is given as her home address. There is no record of Susie returning to the printing trade. Her father, James Moran, was listed in SF Directories as a compositor. Susan appears in no SF Directories after 1878 except 1886.

181. Morgan, Miss Mary J. [Work? 1866], Towne & Bacon. SOURCE: Pay Roll Book, Towne & Bacon, 1865–1873.

Note: She was paid $2.50 per day, apparently a normal rate for typesetting in this shop.

182. Morse, Emma L. 656 Harrison St. (SF). Compositor, Bacon & Co. (1) Initiated SF TU, October 1890 (2). SOURCES: (1) 1889, 1890, 1891, 1892 Directories. (2) PUP, October 1890.

183. Mower, Miss Etta. 1423 Third St. (O) (1). Compositor, S. W. Raveley (O) (1). Compositor, book & job offices (2). Applicant to SF TU (2). Initiated into SF TU (3). "First lady member to hold office in the [SF] Union" (4). SOURCES: (1) 1883 Directory (2) PUP, June 1889. (3) PUP, July 1889. (4) PUP, November 1889.

Note: PUP, November 1889, reported Etta Mower and Nellie E. Fairchild (q.v.) as elected to SF TU membership committee, the first women officeholders in the SF TU.

184. Murdoch, Elizabeth (2). Libbie (1). Compositor, WCPU (1). Printer (2). Born: New York (2). Age: 20 (2). SOURCES: (1) 1869, 1870 Directories. (2) 1870 Census (O).

Note: Boarded with City Marshal Perry Johnson & family (Census). No residence is given in either Directory.

185. Murphy, Delia. San Francisco. Printer. SOURCE: *The History of Woman Suffrage* (Stanton et al., eds., 1877), v. 3: 762.

Note: 1881 Directory: Delia Murphy, res. 1601 Gough, "with The Pacific Trader's Agency" (not a publication). [Cor]delia Marion Murphy married Octavius Roy Dearing, inventor of the two-third California Job Case, in SF on 11 January 1882. She went to Portland, Oregon, with him where a daughter was born 29 June 1896. She returned to Santa Clara County after this date and died on 16 July 1932. The Oak Hill Cemetery in San Jose, where she was interred on 18 July 1932, has no records. The obituary in the

San Jose Mercury Herald (17 July 1932, p. 24:1) reported she was 75 years old. Her daughter was Mrs. Dorothy Marion Britton. There is no further trace of Delia Murphy in any SF, O or SJ Directory. (California State Death Certificate #40320, 1932.)

186. Murphy, Katie. 318 Oak St. (SF). Compositor, WCPU. SOURCE: 1872 Directory.

187. Murray, Mary. Hayes St. (SF) {1}. 118 Hayes St. {2}. Works in Printing Office {1}. Compositor, A. L. Bancroft & Co. {2}. Born: California {1}. Age: 16 {1}. SOURCES: {1} 1880 Census. {2} 1882, 1883 Directories.

188. Myers, Miss Kate Typo with "sits" on the *Tulare Times*. SOURCE: *Merced Express*, 10 July 1880, p. 3:2.

Note: Name is from a news item: "Two young lady typos who have 'sits' on the *Tulare Times*, Misses Kate and Mary Myers, spent the 5th in Merced among relatives." This reference was furnished by Librarian R. Dean Galloway.

189. Myers, Miss Mary Typo with "sits" on the *Tulare Times*. SOURCE: *Merced Express,* 10 July 1880, p. 3:2.

Note: Name is from a news item: "Two young lady typos who have 'sits' on the *Tulare Times*, Misses Kate and Mary Myers, spent the 5th in Merced among relatives." This reference was furnished by Librarian R. Dean Galloway.

190. Nassa, Catherine. Natoma St. (SF). Printer. Age: 16. SOURCES: 1880 Census.

Note: Mother, Mary, kept house at Natoma St. address. See Elizabeth O'Brien.

191. Nauber, Mary. 1312 Larkin St. (SF) {1} {3}. News Compositor {1}. Francis Valentine & Co. {2} Printer, F. F. & Co. {3}. SOURCES: {1} 1888 Directory. {2} PUP, August 1890. {3} 1891 Directory.

192. O'Brien, Elizabeth. Boarding house, Natoma St. (SF). Printer. Born: Ireland. Age 21. SOURCE: 1870 Census.

Note: See Catherine Nassa.

193. O'Grady, Nellie. 424 1/2 Clementina St. (SF). Compositor, Woman's Publishing Co. SOURCE: 1875 Directory.

194. O'Hara, Cassie. [5 1/2] Raush St. (SF). Works in Type Foundry. Born: California. Age: 21. SOURCE: 1880 Census.

Note: See Elizabeth O'Hara and Mary O'Hara.

195. O'Hara, Elizabeth {1}. Miss Lizzie {2} 5 1/2 Raush St.

(SF) (2). Works in Type Foundry (1). Tyepesetter, Painter &
Co. (2). Born: California (1). Age: 19 (1). SOURCES: (1) 1880
Census. (2) 1880 Directory.

Note: See also Cassie O'Hara and Mary O'Hara. Painter & Co. were typefounders.

196. O'Hara, Miss Mary. 5 1/2 Raush St. (SF). Typesetter,
Painter & Co. SOURCE: 1880 Directory.

Note: See also Cassie O'Hara and Elizabeth O'Hara.

197. Outman [?], Emma. Works in Type Foundry. Born:
New York. Age: 14. SOURCE: 1870 Census.

Note: Father, Isaac, shown as laborer in 1870 Census. Name is questionable
because of bad handwriting in Census.

198. Owen, Miss Elma. NE corner Autumn & San Augustin
Sts. (SJ) (1). 211 Autumn St. (SJ) (2). Compositor, *San Jose Mercu-
ry* (1)(2). SOURCES: (1) 1876 Directory (SJ). (2) 1878 Directory
(SJ).

Note: 1876 Directory (SJ): Clifford J. at 211 Autumn; also Charles P. Owens
(sic). 1878 Directory (SJ): C. J. and Charles P. Owen are at 211 Autumn and
listed as printers. Miss Lucy is listed as a student. See Jennie Owen.

199. Owen, Miss Jennie. 211 Autumn St. (SJ). Printer.
SOURCE: 1878 Directory (SJ).

Note: Father was J. J. Owen, editor & proprietor of the *San Jose Daily and
Weekly Mercury*, at same address. 1874 Directory (SJ) shows D. D. and C. J.
Owen as printers at *Mercury.* D. D. lived on Alameda Road, C. J. at 211 Au-
tumn. See Elma Owen.

200. Ownby, Miss Annie M. 69 N. Whitney St. (SJ). Printer,
San Jose Daily Herald. SOURCE: 1887 Ulhorn & McKenney Directory (SJ).

201. Parker, Mary. 725 Tehama (SF) (1). Compositor (1).
Apprentice typesetter, *Pacific Monthly.* (2) Age: 18 (2)
SOURCES: (1) 1864 Directory. (2) *Pacific Monthly,* August 1864,
p. 825.

Note: Mary Parker was a founding member of the Female Typographical
Union. She was killed 5 August 1964 in an accident while visiting Green
Valley, Sonoma County, according to Source (2)

202. Patrick, Jennie E. Oakland (2). 212 Powell St. (SF) (4).
703 Geary St. (5). Hotel Miramar [SW corner Geary & Larkin
Sts.] (6)(7). Book & job printer (1)(2). Herald of Trade Pub.
Co. (3). D. F. Cooper & Co., props. of Herald of Industry Trade
Book & Job Printing Office [Jennie E. Patrick] (4). Printer,

(5)(6). Mgr., Illustrated Pub. Co. (7) Business addresses: 429 Montgomery St. (1). 415 Market St. (3). 413-415 Market St. (4). 6 Mills Building, 3rd Floor (7). SOURCES: (1) 1887 Directory. (2) 1888 Directory. (3) 1890 Directory (4) 1891 Directory. (5) 1893 Directory. (6) 1894. (7) 1895 Directory.

Note: D. F Cooper & Co. (D. F. Cooper, Jennie E. Patrick and George W. Bumm), were publishers of the *California Dramatic News* (weekly) in 1891 Directory. Not found in Oakland Directories and not in 1892 SF Directory nor after 1895. 1896 Directory names principals of B. H. Patrick Co., office supplies and printing, but Jennie E. Patrick not listed. The California Historical Society Library has two billheads with Jennie E. Patrick imprints.

203. Payne, Mrs. W. E. Compositor, book & job offices. SOURCE: PUP, February 1889.

204. Pendegast, Lizzie. Third St. (SF). Typesetter. Born: Mississippi. Age: 20. SOURCE: 1880 Census.

205. Peterson, Mrs. Agnes B. 811 Stockton St. (SF) (2). Compositor (1). Superintendent, WCPU, 517 Clay St. (2). SOURCES: (1) *The Revolution*, 10 September 1868, p. 149:2. (2) 1868 Directory.

Note: *The Daily Examiner*, 9 July 1868, lists "Mrs. Peterson and child" as having arrived on the P.M.S.S. *Constitution* at 11 a.m. that day. This coincides with the time frame of her arrival mentioned in a letter to *The Revolution*, dated 14 August. Named Mrs. Agnes B. Peterson Cotton in SF *Mercury*, 19 June 1869, p. 1:3.

206. Pickard, Sarah. 26 Minna St. (SF). Compositor, WCPU. SOURCE: 1872 Directory.

207. Pitts, Miss E. A. (1). Pitts, Mrs. E. A. (2)(3). 122 Taylor St. (SF) (1). 420 Montgomery St. (3). Teacher, Calisthenics House (1). Proprietor (with Frank Wickes), *Saturday Evening Mercury* (3). SOURCES: (1) 1867 Directory. (2) 1869 Directory. (3) 1870 Directory.

Note: She became Mrs. Emily Pitts Stevens (q.v.), a founder of the Women's Co-operative Printing Union and one of the most active women in the suffragist cause. She also appears as a teacher in the San Francisco Municipal Reports for 1866–67 and 1867–68, the first public records of her in the city.

208. Pitts Stevens, Mrs. Emily (3)(4)(5)(6). Stevens, Mrs. Emily Pitts (1)(2)(9). Stevens, Emily P. (7). Stevens, Emily Pitts (8). 538 Greenwich St. (SF) (1). 607 Pine St. (2). 1327 Broadway (3). 1513 Larkin St. (4)(5). 2194 Divisadero St. (6)(7)(8). 1907 Bush St. (9). Journalist (3)(5)(6). Temperance Lecturer (8)(9). A

founder of the WCPU; trainer of female type-setters (10)(11).
SOURCES: (1) 1871, 1872 Directories. (2) 1873 Directory. (3)
1875 Directory. (4) 1876 Directory. (5) 1876, 1877 Directories.
(6) 1878 Directory. (7) 1879 Directory. (8) 1880 Directory. (9)
1881 Directory. (10) San Francisco *Call*, 1 November 1891, p.
12:3. (11) Advertisement, *The Pioneer*, 19 December 1872, p. 6:4

Note: An article in *The Pioneer* (29 August 1872, p. 8:1) describes her early experiences as a compositor. From 1882 on, she is not listed in any Directory but her husband continues at the 1907 Bush St. address through 1907. She was also founder, president and superintendent of the Woman's Publishing Company which she managed until it was sold (February 1875) to Amanda Slocum (q.v.). Her ad in *The University Echo* also said she was opening a "School of instruction for Girls and Women, who are dependent and wish to learn the art of Type-setting and Printing as a means of livelihood." See also Emily Pitts.

209. Poggi, Minnie. 430 Vallejo St. (SF) (1). Compositor, Dewey & Co. (1). Compositor (2). Born: California (2). Age: 20 (2). SOURCES: (1) 1880 Directory. (2) 1880 Census.

210. Preston, Albertine. 861 Folsom St. (SF). Compositor, WCPU. SOURCE: 1872 Directory.

211. Preston, Clara. Compositor, Dewey & Co. (1) Works at type setting (2). Compositor, Bacon & Co. (3). Born: California (2). Age: 16 (2). SOURCES: (1) 1880 Directory. (2) 1880 Census. (3) 1881 Directory.

212. Preston, Miss Emma. 730 Sixteenth St. (O). Compositor, Hollis & Carleton. SOURCE: 1885 Directory (O).

Note: Miss Emma Preston shown as dressmaker, same address, in 1886 Directory (O). See Nellie Preston.

213. Preston, Nellie. 730 Sixteenth St. (O). Compositor. SOURCE: 1885 Directory (O).

Note: See Emma Preston.

214. Punch, Bridget. Davis St. ("boarder") (SF). Typesetter. Born: Connecticut. Age: 20. SOURCE: 1880 Census.

215. Ravenscroft, Mrs. Minnie. 1110 Montgomery St. (SF) (1). 1224 Filbert St. (3). Compositor, *Bulletin* (1) (3). Compositor, Bacon & Co., *Bulletin*, Dewey & Co., and *Rural Press* (2). SOURCES: (1) 1888 Directory. (2) BLS Inv. (3) 1889, 1890 Directories.

Note: 1886, 1887 Directory: Dewey A. Ravenscroft, compositor, *Bulletin*, res. 1110 Montgomery St. 1888 Directory: Dewey and Minnie listed at same address and as compositors, *Bulletin.* 1889 Directory: both listed as at 1224 Filbert St. Minnie's testimony in BLS Inv., p. 9. From the record, this might have been a husband-wife situation, although Dewey could have been another relative with same data.

216. Reed, Lizzie. 525 1/2 Howard St. (SF). Compositor, Woman's Publishing Co. SOURCE: 1875 Directory.

217. Rerat, Miss Louise. Louisa (3). 866 Willow St. (O) (1). 371 10th St. (2). Willamette House (3). 819 Harrison St. (5). Compositor, *Times* (1). Printer, Job Dept., Tribune Pub. Co. (2). Printer, *Morning Times* (3). Compositor, *Times* (formerly, Pacific Press) (4). Compositor, *Oakland Tribune* (5). SOURCES: (1) 1886 Directory (O). (2) 1887 Directory (O). (3) 1888 Directory (O). (4) BLS Inv. (O). (5) 1889 Directory (O).

Note: Testimony in BLS Inv. (O), p. 16.

218. Richmond, E. N side William St. near Telegraph Ave. (O). Widow. SOURCE: 1875 Bishop Directory (O).

Note: This is actually Mrs. Lizzie G. Richmond of the WCPU (q.v.). A William Richmond, printer, is listed at same address. He was, in fact, Willard P. Richmond. See the next two entries under Richmond. The designation "Widow" is not accurate here and sometimes may have been used in Directories to mean "grass widow," i.e., a divorcee, which was true in this instance.

219. Richmond, Mrs. Lizzie G. (1)(2)(3)(4)(7)(8)(9)(10). Richmond, Mrs. L. G. (11)(12). Richmond, E. (6). Richmond-Judd, Mrs. Lizzie G. (13)(14)(15)(16)(18)(19)(20) (22)(24). Richmond-Judd, Mrs. L. G. (17)(21)(23)(25) (26)(27). Judd, Mrs. Lizzie G. (28)(29)(32). Judd, Mrs. L. G. (30)(31). 617 Mason St. (SF) (1). 520 Sutter St. (2). 726 California St. (3). 603 Ellis St. (4). N side William near Telegraph (O) (6)(8)(9)(12). 518 William St. (O) (14)(16) (18)(19)(26)(28)(29)(30)(31)(32). The Baldwin Hotel (SF) (21)(22). 711 Jones St. (23)(24)(25). Oakland (7)(10)(11)(13)(15)(20)(27). Manager, WCPU (1)(2)(3)(4)(7). Superintendent, WCPU (5). Widow (6). Printer (7)(16)(17) (18). Proprietress, WCPU (10)(11)(13)(14). Printer, WCPU (12). (Mrs.) L. G. Richmond & Son (15)(19)(20)(24)(26) (27). Proprietor, with Willard P. Richmond, WCPU (21)(22) (23). Vice-President, Hicks-Judd Co. (28). Age: 42 (17). Born: Rhode Island (17). SOURCES: (1) 1869, 1870 Directories. (2) 1871 Directory. (3) 1872, 1873 Directories. (4) 1875 Directory. (5) 1875 Bishop Directory. (6) 1875 Bishop Directory (O). (7) 1876 Directory. (8) 1876 Directory (O). (9) 1877 Directory. (10)

1877 Directory. (11) 1877, 1878 Bishop Directories. (12) 1878 Directory (O). (13) 1879 Directory. (14) 1879 Bishop Directory (O). (15) 1880 Directory. (16) 1880 Directory (O). (17) 1880 Census. (18) 1880 Directory (O). (19) 1881 Directory (O). (20) 1882 Directory. (21) 1883 Directory. (22) 1884 Directory. (23) 1885 Directory. (24) 1886 Directory. (25) 1887 Directory. (26) 1887 Directory (O). (27) 1888 Directory. (28) 1891 Directory (O). (29) 1892, 1893 Directories (O). (30) 1894 Directory (O). (31) 1895 Directory (O). (32) 1896 Directory (O).

Note: In later years, e.g. 1885 and 1886, Mrs. Judd had herself listed three ways in the San Francisco Directory: Judd, Mrs. L. G.; Richmond-Judd, Mrs. Lizzie G.; and Richmond, Mrs. L. G. & Son (Mrs. L. G. Judd & Willard P. Richmond, proprietors, Women's Co-operative Printing Union). These covered her name changes and assured publicity for her firm. Nelson A. Judd, her husband, remained in Oakland at the 518 William St. address through 1895. By 1897 he was listed as living at the Hotel St. Nicholas in San Franciso. From 1891 he was always listed as President of the Hicks-Judd Co.

220. Richmond, Mrs. Mary H. (Shearer). 514 Stockton St. (SF). W. P. Richmond & Co, Job Printers (1). SOURCE: (1) 1893 Directory.

Note: Home and business were at 514 Stockton St. 1894 Directory has printing-office only at this address. This enterprise evidently was begun after her husband, Willard P. (q.v.), left Hicks-Judd, successors to WCPU. She died 31 December 1942.

221. Robins, Mrs. Emily. 1 Taylor St. (SF) (1). 605 Pine St. (2). Superintendent, Woman's Publishing Co. (511 Sacramento St.) (1)(3)(4). SOURCES: (1) 1874 Directory. (2) 1875 Bishop Directory. (3) Advertisement, passim, *The Pioneer*, October 1873. (4) *The University Echo*, April 1873, p. 8:1.

Note: The ad cited originally appeared with Emily Pitts Stevens as president and superintendent and both ads contained the same statement that there was a "School of instruction for Girls and Women, who are dependent and wish to learn the art of Type-setting and Printing as a means of livelihood." An Emily Robins appears in the 1870 Census, age 25, born in Maine. Her listed occupation is almost illegible but appears to have been sewing. (1870 Census, San Francisco, ward 9, p. 144, line 33.)

222. Robinson, Jewell. 1204 Folsom St. (SF). Printer. SOURCE: 1889 Directory.

Note: Not in 1886–1888 Directories. A Colfax J. Robinson, compositor, J. R. Brodie & Co., also at 1204 Folsom St. in 1889.

223. Robinson, Miss Josie. Card deposited from San Jose TU #231 (1). Withdrew from SF TU (2). SOURCES: (1) PUP, November. 1888. (2) PUP. December 1888.

Note: "J." on card deposited.

224. Roche, Miss. [Work? 1868], Towne & Bacon. SOURCE: Pay Roll Book, Towne & Bacon, 1865–1873.

Note: Entry and signature for pay are always "Miss Roche," nothing more. She was one of the lower-paid women @ $1.33 1/3 per day.

225. Roney, Mrs. F. G. [Mrs. Frank G.] [254 Folsom (SF)] Exempt member, SF TU. SOURCE: PUP, June 1889.

Note: Frank G. Roney, compositor, *Alta*, at 254 Folsom in 1889 Directory, but Mrs. Roney not listed.

226. Roper, Miss M. E. Harris House [59 & 65 Pacific Ave.], Santa Cruz. Printer. SOURCE: 1887–88 Ulhorn & McKenney Directory (SJ).

227. Ross, Miss Hattie Compositor, *Examiner*. SOURCE: PUP, October. 1888.

228. Ross, Miss Hattie (2)(3). Miss H. (1). Compositor, *Examiner* (1)(3). Card deposited from Multnomah TU [Portland, Oregon] #58 (2). SOURCES: (1) 1889 Directory. (2) PUP, October 1888. (3) PUP, October 1888–February 1889.

229. Rue, Mrs. H. P. 717 Twelfth St. (O) (1). Compositor, Pacific Press (1). "I have charge of the lady apprentices in the type-setting department of Pacific Press... I make up forms for stereotyping, etc." (2). SOURCES: (1) 1886, 1887 Directories (O). (2) BLS Inv. (O).

Note: Testimony in BLS Inv. (O), p. 21. A Harry P. Rue, listed as a printer at Pacific Press, appears at same address in Directories cited.

230. Ryan, Miss M. Exempt member, SF TU. SOURCE: PUP, October 1888.

Note: See May Ryan. PUP shows her as exempt member, SF TU, from October 1888–June 1889.

231. Ryan, May. Compositor, *Alta*. SOURCE: 1888 Directory.

Note: See Miss M. Ryan.

232. Schlesinger, Mrs. Julia (1). 32 Ellis St. (SF) (1). Journalist (1). Editor and co-publisher, *Carrier Dove*, Oakland and San Francisco, 1883–1893 (2). Co-proprietor, Carrier Dove Printing and Publishing Co., from February 1888 (3). Supporter of suffrage, woman's rights and employer of woman typesetters

(4). SOURCES: (1) 1889 Directory. (2) Masthead, *Carrier Dove*, passim. (3) *Carrier Dove*, 25 February 1888, p. 139:3 and 5 March 1888, p. 150:1–2. (4) *Carrier Dove*, passim, and interview with great-great-great grandson.

Note: See also Nellie Gorman, who became informant Michael Engh's great-great grandmother after she married Milo H. Fish, son of Julia Schlesinger by a previous marriage.

233. Schneider, Miss Minnie. 24 Erie St. (SF)(1). Compositor(1). Initiated into TU, November (2). SOURCES: (1) 1886 Directory. (2) PUP, November 1890.

234. Schneider, Wilhelmina. 24 Erie St. (SF) (2). Compositor (1) (2). Born: California (1). Age: 17 (1). SOURCES: (1) 1880 Census. (2) 1880, 1881 Directories.

235. Scholl, Mrs. N. J. Exempt member, SF TU. SOURCE: PUP, December 1889.

Note: 1890 Directory (SF) lists an Albert J. Scholl, compositor, *Call*, residing at 678 Harrison St. (SF).

236. Scully, Miss. [Work? 1869], Towne & Bacon. SOURCE: Pay Roll Book, Towne & Bacon, 1865–1873.

Note: Entry and signature for pay are always "Miss Scully," nothing more. She was one of the lower-paid women @ $1.66 2/3 per day.

237. Seiler, Miss Clara. 447 Minna St. (SF) (1). Compositor (1). Compositor, Filmer-Rollins Electrotype Co. (2). SOURCES: (1) 1886 Directory. (2) PUP November 1890.

238. Sickler, Miss Nellie. 2600 Folsom St. (SF). Printer. SOURCE: 1883 Directory.

239. Slocum, Amanda (1). Amanda M. (4). Mrs. A. M. (2). Mrs. Amanda M. (6). San Jose (1). 236 Montgomery St. (SF) (2). 615 Third St. (3). 5 Ewer Pl. (4)(5). 726 Washington St. (6). Commercial Hotel (6)(7). 612 Clay St. (8). 115 Sutter St. (9). 629 Clay St. (10). 1920 Sacramento St. (11)(12). Superintendent, Woman's Publishing Co. office [605 Montgomery St.] (2)(3). Solicitor, Taylor & Nevin (4). Book, card & job printer [534 Commercial St.] (5). Job printer (6)(7)(8). Book & job printer (9). Printer [543 Clay St.] (10)(11)(12). Born: Iowa (1). Age: 29 (1). SOURCES: (1) 1870 Census. (2) 1875 Bishop Directory. (3) 1876 Bishop Directory. (4) 1877 Directory. (5) 1877 Bishop Directory. (6) 1878 Directory. (7) 1879 Directory. (8) 1880 Directory. (9) 1881 Directory. (10) 1882

Directory. (11) 1883 Directory. (12) 1884 Directory.

Note: Business and home addresses are not always clearly differentiated in Directory entries. 236 Montgomery St. was both home and publishing address for *Common Sense* and the Woman's Publishing Co. when the Slocums acquired that firm and moved it from 511 Sacramento St. The Slocums separated and divorced (he remarried in 1882), and Amanda appears in the 1887 Directory as Mrs. A. M. Reid at the 1920 Sacramento St. address but is unlisted thereafter. However, the surname is spelled Reed in Stanton et al., *History of Woman Suffrage,* v. 3:761, where she is described as having "capacity for business, by the management of a large printing and publishing establishment for many years." The 1875 Directory shows Amanda Slocum's "office" as 507 Montgomery St., under her entry and 605 Montgomery St. under Woman's Publishing Co. listing. See also Amanda Taylor, Clara Eldridge, Clara Slocum and Eva Slocum.

240. Slocum, Clara. San Jose (1)(2). [See Amanda M. Slocum for SF residences.] Compositor, *Roll Call* (3). Born: San Jose, California, 28 February 1863 (1). Age: 7 (2). SOURCES: (1) *The Slocums of America . . .* (Charles Elihu Slocum, 1908, p. 260.) (2) 1870 Census. (3) *History of Woman Suffrage* (Stanton, Anthony et al., eds., 1877), v. 3:762.

Note: A daughter of Amanda M. Slocum (q.v.). "Mrs. Slocum began the publication of *Roll Call,* a temperance magazine, which was mainly edited by her little daughter Clara, only fifteen years old, who also set all the type." "The Enchanted Child, a Fairy Story," appeared in *Sunshine,* "A Monthly Magazine for Young Readers" (May, 1875, pp. 114:1–115:2), "By Carrie Slocum (age 12 years) . . . ." See also entry for Clara Eldridge, her married name.

241. Slocum, Eva T. San Jose (1)(2) 236 Montgomery St. (SF) (3). 5 Ewer Place (4). Printer, *Common Sense* (3). Compositor, Taylor & Nevin (4). Born: Iowa (1). New York (2) Age: 5 months (1); 10 years (2). SOURCES: (1) 1860 Census, Santa Clara County. (2) 1870 Census. (3) 1875 Directory. (4) 1877 Directory.

Note: Eva T. Slocum was a daughter of Amanda M. [Taylor] Slocum's first marriage. She was about two years old when her mother married William N. Slocum. Her name is given as Etta M. in the 1860 Census enumeration for Santa Clara County, taken on 28 July 1860.

242. Smith, Miss A. Exempt member, SF TU. SOURCE: PUP, October 1888.

243. Smith, Mrs. Frank D. Card deposited from Spokane Falls [Washington] #193. Withdrew card from SF TU.

SOURCE: PUP, March 1889.

244. Smith, Miss Kitty E. 1816 Market St. (SF). Compositor. SOURCE: 1883 Directory.

245. Smith, Margaret (1). Miss Margaret (4)(7). Miss Maggie (2)(5)(6)(8)(9). M (3). 3 Turk St. (SF) (2). 521 Howard St. (3). 1410 Sutter St. (4). 109 Tehama St. (5) 112 Taylor St. (6). 2009 Powell St. (7). 1834 Powell St. (8)(9). Compositor (1). Compositor, *Saturday Evening Mercury* (2). Compositor, *Morning Call* (3)(4)(5)(6)(7)(8)(9]. Born: California (1). Age: 15 (1). SOURCES: (1) 1870 Census. (2) 1870 Directory. (3) 1871 Directory. (4) 1873 Directory. (5) 1875 Bishop Directory. (6) 1876 Directory. (7) 1878 Bishop Directory. (8) 1882 Directory. (9) 1884, 1885, 1886 Directories.

Note: Maggie Smith is listed at 317 Vallejo St. in 1887 Directory, but this is an error because Mary Smith (q.v.) had been listed at this address in 1883 Directory, and Maggie had been at 1834 Powell St. on dates before and after this entry. A Miss M. Smith signed a petition to the legislature in behalf of suffrage in 1870. Margaret Smith was then working at the *Saturday Evening Mercury* for Emily Pitts Stevens (q.v.), one of the leaders in the movement, so it is likely that she was canvassed to sign.

246. Smith, Miss Mary. 308 Leavenworth St. (SF). Compositor. SOURCE: 1883 Directory.

247. Smith, Miss Mary E. (1)(3)(6). Miss M. E. (2). Miss M. (5)(8)(9). 13 Verona Pl. (SF) (1)(2). 317 Vallejo St. (3)(4). Compositor (1)(3)(4). Compositor, *Jolly Giant* [weekly] (2). Compositor, *Call* (5)(6)(8). Printer (7). SOURCES: (1) 1876 Bishop Directory. (2) 1876 Directory. (3) 1883 Directory. (4) 1887 Directory. (5) 1888, 1889 Directories. (6) 1890 Directory. (7) 1891 Directory. (8) PUP, October 1888–December 1889.

Note: The 1887 Directory lists Maggie Smith (q.v.) at 317 Vallejo St., but this is an error because on both sides of this date Maggie Smith was at 1834 Powell St.

248. Smith, Mrs. Mate A. 520 Pine St. (SF). Compositor, *Daily Echo.* SOURCE: 1878 Directory.

249. Smith, Minnie. 1108 Powell St. (SF). Press feeder. SOURCE: 1892 Directory.

250. Somers, Miss Minnie L. 1017 Clay St. (O) (1). Compositor, Pacific Press (1)(2). SOURCES: (1) 1891 Directory (O). (2) BLS Inv. (O).

251. Sorrels, Laura. Card deposited from Fresno City TU

#144 (1). Card issued, SF TU (2). SOURCES: (1) PUP, September 1888. (2) PUP, October 1888.

252. Spooner, Ida B. 1005 Powell St. (SF). "Works in Printing Office." Born: Massachusetts. Age: 20. SOURCE: 1880 Census.

253. Stacey, Emma. [404 Geary St. (SF).] Printer. Born: Pennsylvania. Age: 18. SOURCE: 1880 Census.

Note: Her father, Marcus Stacy, 54, is listed as a printer in Census. Name spelled Stacy in Census but three Directories show Mary, her mother, as Stacey (q.v.).

254. Stacey, Mrs. Mary N. (1)(3). Mary (2). Mary N., widow (4) 404 Geary St. (SF) (1)(3). 812 Howard St. (4). Compositor, *Alta* (1)(3). Printer (2). Compositor (4). Born: Pennsylvania (2). Age: 50 (2). SOURCES: (1) 1880 Directory. (2) 1880 Census. (3) 1881 Directory. (4) 1882 Directory.

Note: Her husband, Marcus Stacy, 54, listed as a printer in Census. Family name given as Stacy in Census but three Directories list Mary as Stacey. See Emma Stacey.

255. Stevens, Mrs. Mary. Compositor, WCPU. SOURCE: 1871 Directory.

256. Stevenson, Miss Nellie. 1065 Fifth Ave. (O). Compositor, *Christian Independent*. SOURCE: 1888 Directory (O).

257. Stewart, Miss Maggie. 289 Union St. (SF). Compositor. SOURCES: 1882, 1883 Directories.

258. Stickney, Mrs. E. R. Honorary member, SF TU. SOURCE: PUP, October 1888.

259. Still, Miss Ada. 506 E. Fifteenth St. (O). Compositor, *McMullin's New Weekly*. SOURCE: 1888 Directory (O).

260. Stockton, Annie E. Tenth, between William & San Salvador Sts. (SJ). Compositor, *Argus* Office. SOURCE: 1878 Directory (SJ).

261. Stoner, Miss C. Corner Washington & Liberty Sts. (SJ). Printer, *San Jose Daily Herald*. SOURCE: 1887 Ulhorn & McKenney Directory (SJ).

262. Stow, Marietta (1)(2). Marietta L. (7). Mrs. J. W. (3). Lizzie (1). 2 Eddy St. (SF) (1). 304 Stockton St. (3). 1921 Sacramento St. (4)(5)(6)(7). Authoress (1). Compositor, *Woman's Herald of Industry* (2). Editor, *Woman's Herald of Industry* (4). Authoress & publisher (5). Journalist (6). Widow (7). Born:

Massachusetts (1). Age: 45 (1). SOURCES: (1) 1880 Census. (2) *Woman's Herald of Industry*, October 1881, p. 2:1. (3) 1882 Directory. (4) 1884 Directory. (5) 1885 Directory. (6) 1886 Directory. (7) 1888 Directory.

Note: See Mrs. Laura W. Briggs. Mrs. Stow is not listed in Directories omitted in sequence. Her married name was Mrs. Joseph W. Stow.

263. Summers, Miss Helen (1)(2). Miss Elna (3). 717 Twentieth St. (O) (1)(2)(3). Compositor, Times Publishing Co. (1). Compositor, *Enquirer* (2). Printer, *Enquirer* (3). SOURCES: (1) 1886 Directory (O). (2) 1887 Directory (O). (3) 1888 McKenney Directory (O).

Note: See Miss Nellie Summers.

264. Summers, Miss Nellie. 717 Twentieth St. (O) (1)(2). Compositor, Times Publishing Co. (1)(2). Compositor, *Enquirer* ("formerly, Pacific Press") (3). SOURCES: (1) 1884 McKenney Directory (O). (2) 1886–87 Directory (O). (3) BLS Inv. (O).

Note: Testimony in BLS Inv. (O), p. 17. See Miss Helen Summers.

265. Sweeney, Miss Mamie (1). Miss M. (3)(5). 1505 Post St. (SF) (1). Compositor, Dewey & Co.; Compositor, *Daily Journal of Commerce* (1). Employed on *Daily Journal of Commerce* (2)(4). Printer, *Bulletin* (5). SOURCES: (1) 1889, 1890 Directories. (2) BLS Inv.. (3) PUP, October 1888. (4) *Daily Journal of Commerce* roster, PUP 1888–89. (5) 1891 Directory.

Note: Testimony in BLS Inv., p. 8. See Mary E. Sweeney. The 1505 Post St. address may have been an error for 1505 Pine St. where the family lived as late as 1897.

266. Sweeney, Mary E. 1505 Pine St. (SF) (2). Apprenticed to Printer (1). Born: California (1). Age: 17 (1). SOURCE: (1) 1880 Census. (2) 1880 Directory.

Note: See Miss Mamie Sweeney. Differentiating the Sweeneys is difficult because of the use of nicknames and the mixup in street addresses.

267. Taylor, Amanda. Corner 23rd & Alabama Sts. (SF). Compositor, *Occident* (1). Editor and typesetter, *The Olive Branch* (2). SOURCES: (1) 1875 Directory. (2) *The Olive Branch*, Vol. 1, #1, October 1873, 2:2.

Note: The first female in SF to enter amateur journalism. "Miss Amanda Taylor deserves great credit for attempting so much and should have the earnest support of all. None who see *The Olive Branch* can be otherwise than pleased with its neat appearance and good common sense articles, while those who read its editorials and think that it is all the work of one of 'California's fair daughters' feel justly proud of her and the state which produces such girls."(*Pacific Monthly*, November 1873, p. 52)

268. Taylor, Amanda (1) Amanda M. (2). San Jose (1). Born: Iowa (1). Age: 20 (1). SOURCES: (1) 1860 Census, Santa Clara County. (2) *The Slocums of America*...(Charles Elihu Slocum, 1908), p. 260.

Note: According to the Slocum history, p. 260, "...[William N. Slocum] married first in 1862 widow Amanda M. Taylor of San Francisco" and in addition to daughter Clara (q.v.), there was a son, Frederick, born February 1864, who died in October 1868. The 1860 Census shows her living in San Jose with an Anderson family, all from Iowa. The female head of the household is listed as "H. A." Anderson; the other Andersons were aged 25, 16 and 12. Amanda was 20 at this time, fits into sibling age-spread and thus might have also been an Anderson.

269. Taylor, Miss Lillie. 768 Fifth St. (O) (1)(2). 1302 San Pablo Ave. (O) (6). Compositor, *Oakland Times* (1)(2). Compositor, *Enquirer* (3)(5). Printer, *Enquirer* (4) Compositor, *Tribune* (6). SOURCES: (1) 1884-85 Bishop Directory (O). (2) 1886 Directory (O). (3) 1887 Directory (O). (4) 1888 Directory (O). (5) BLS Inv. (O). (6) 1892 Directory.

Note: Testimony in BLS Inv. (O), p. 17.

270. Terry, Hattie. Polk St. (SF). Printeress. Born: California. Age: 19. SOURCE: 1880 Census.

271. Thompson, Ida. 321 1/2 Union St. (SF). Compositor, WCPU. SOURCE: 1872 Directory.

272. Tory, Harriet. Apprentice Compositor. Born: England. Age: 16. SOURCE: 1870 Census.

273. Upton, Augustine F. (1)(2)(3). A. F. (4). Compositor, Bacon & Co. (1) Compositor, V. A. Torras & Co. (2). Compositor, Valentine & Co. (3) Printer at Dewey's (4). SOURCES: (1) 1885 Directory. (2) 1886 Directory. (3) 1889 Directory. (4) BLS Inv.

Note: Left Bacon after 18 years because Bacon said he could procure men to work for $18. per week (BLS Inv.).

274. Way, Mrs. Marjorie. [Work? 1868], Whitten & Towne. SOURCE: Whitten & Towne Receipt Book, Vol. 26, Towne & Bacon Records, Stanford University.

Note: Wages of $20.00, shown for week ending 14 August 1858, are comparable to those paid the men who are definitely identifiable as printers after leaving Whitten & Towne, although this Receipt Book gives no clues to work done. An entry in the Cash Book likewise gives no indication of specific tasks but the pay scale indicates typesetting.

275. Weeks, Miss Lucy R. 1178 Regent St. (A). Compositor, *Alameda Encinal.* SOURCE: 1890 Directory (O).

276. Welsh, Maggie. SW corner Baker & Bush Sts. (SF) {1}. Compositor, Bacon & Co. {2}. Compositor, WCPU {2}. SOURCES: {1} 1888 Directory. {2} BLS Inv.

Note: Testimony in BLS Inv., p. 9.

277. Wheeler, Mrs. Louise M. Member, SF TU, June 1876, by transfer from Washoe (Nevada) #65. SOURCE: *Gold Hill* [Nevada] *News*, June 1876.

Note: Listed from Robert D. Armstrong, *Nevada Printing History . . . 1858–1880* (Reno: University of Nevada Press, 1981), p. 279. The *News* reprinted the item from the *Carson Appeal:* "The Typographical Union of San Francisco, at its meeting yesterday, admitted to membership Louise M. Wheeler, who presented a card from Washoe Union. This is the first lady admitted into the San Francisco Union, and makes quite a departure from the old rule." The *News* commented: "Mrs. Wheeler is well known in Gold Hill, having worked for several weeks on the *News.*"

278. Williams, Flora. Montgomery St. (SF). Printer. Born: California. Age: 16. SOURCE: 1880 Census.

Note: Brother, David, also shown as a printer in 1880 Census.

279. Wilson, Minnie. 13th St. (SF). Compositor, John H. Carmany & Co. SOURCE: 1875 Directory.

280. Winters, Miss Helen C. 735 Harrison St. (O) {1}. Compositor, Bacon & Co. (SF) {1}. Compositor, Dewey & Co. (SF) {2}. SOURCES: {1} 1885 Directory (O). {2} BLS Inv.

Note: Testimony in BLS Inv., p. 12.

281. Wood, Miss Helen T. {1}. Helen {2}. 825 23rd St. (O) {1}. Compositor, Pacific Press {1}{2}. SOURCES: {1} 1886, 1887, 1888 Directories (O). {2} BLS Inv. (O).

Note: Testimony in BLS Inv. (O), p. 21.

282. Woodeson, Miss Alice {1}. Miss Alice C. {2}. 511 M St. (Sacto.) {2} Initiated into membership, Sacramento TU #46 {1}. Employee, *Bee* office {2}. SOURCES: {1} Minutes of Sacramento TU #46, v. 2:121 (25 March 1888). {2} 1889 Sacramento Directory.

Note: Minutes of this Union are in The Bancroft Library. She applied in January, the application was held over one month, and then she was elected and initiated, the first woman member of Sacramento TU #46.

APPENDIX B

The following checklist is intended to give an overview of the various kinds of printing handled by the woman-run firms but does not include the very wide range of ephemera, such as billheads, posters and the like, which were mainstays of their job work.

Pagination in wrappered imprints never includes the blank which is conjugate with a paged leaf. The legal briefs all have wrappers and some have title pages as well as cover titles. The Women's Co-operative Printing Union (and its variant imprints) frequently used numerals to indicate signatures.

The longest hardbound book in the listing, *Whisperings* (WCPU, B 72), has signatures and another characteristic seen often: the blank, half-title leaf is not integral with the rest of the book but is tipped to the free endsheet. *Whisperings* has gatherings of eight leaves (octavo) and measures 6 $^1/_2$" by 5".

Most WCPU books and pamphlets appear to have been (side) stabbed. A few smaller pamphlets are sewn through the center of the single gathering, e.g. B 76.

The binding styles of the hardbound books are typical of their time and place and comparable to the work done by competitors. There is no direct evidence that any of the woman-run shops included a bindery other than the simple tools required for stabbing and sewing.

Amanda M. Slocum announced the purchase of the Woman's [Pacific Coast] Publishing Co. in February 1875 and removed it from 511 Sacramento St. to 605 Montgomery St., former location of the *Evening Post*. She used her own variants of the firm's corporate name in her imprints briefly thereafter. An atypical example is Woman's Book and Job Printing Association (B 36). (See Imprints listing below.)

An adaptation of the standard LC style is used for all entries. Location of copies uses LC designations for libraries.

No attempt has been made to make the legal briefs consistent to one style; most were transcribed from catalog cards.

IMPRINTS OF WOMEN'S PRINTING-OFFICES
1857–1890

Women's Co-operative Printing Union
Women's Book and Job Printing Office
Women's Co-op. Printing Office
Women's Co-operative Print
Women's Co-operative Printing Office
Women's Co-operative Printing Union
Women's Co-operative Printing Works
Women's Co-operative Union [Not to be confused with another non-printing firm.]
Women's Co-operative Union Print
Women's Print
Women's Printing Office
Women's Printing Union
Women's Union Book and Job Printing Office
Women's Union Print
Women's Union Printing Office
Women's Pacific Coast Publishing Co. [Emily Pitts Stevens]
Law Printing House of Woman's Publishing Co.
Woman's Publishing Co.
Woman Publishing Co.'s Print

Amanda M. Slocum
A. M. Slocum, Book and Job Printer
A. M. Slocum, Printer
Amanda M. Slocum, Book and Job Printer
Amanda M. Slocum, Printer
Woman's Book and Job Printing Association
Woman's Printing Association

Note: Amanda M. Slocum and her husband, William N., were publishers and editors of *Common Sense*, a weekly journal, when they acquired the Woman's [Pacific Coast] Publishing Co. from Emily Pitts Stevens and moved it to 605 Montgomery St. (from 511 Sacramento St.) in February 1875. To establish the new identity, they used Woman's Printing Association as their imprint. Following the closing of their publication in June 1875 and a separation, Mrs. Slocum worked elsewhere with her daughter and when she resumed printing on her own in 1877, she never used any imprints which did not contain her name, as listed above. Thus the Woman's Printing Association imprint and its variant appeared only very briefly (see Chapter 8, part I). Generally, items with 511 Sacramento St. in the imprint were under the control of Emily Pitts Stevens or Emily Robins. Everything from 605 Montogomery St. was an Amanda Slocum production.

MISCELLANEOUS IMPRINTS

Folsom Telegraph Newspaper and Job Printing Office (Mrs. John L. Howe, Jr., 1876)
Jennie E. Patrick Printing Co.

CHECKLIST OF WOMEN'S BOOKS AND PAMPHLETS

1. Alaska Coal Company. Prospectus of the Alaska Coal Mining Company. San Francisco: Women's Co-operative Printing Union, 1871, 8 pp. CU-B.
Note: Cover title and title, both set in Caslon caps.; woodcut frontispiece on p. [ii].
2. Amador Coal and Mining Company. Documents and Reports Relating to Amador Canal and Mining Company . . . .

San Francisco: Women's Union Print, 1873, 8 pp. CLU.

3. Anthony, Mark. In the Supreme Court . . . California . . . Respondent . . . . San Francisco: Women's Co-operative Union Print, 1875, 11, 6 pp. CLU.

4. Anthony, Mark et al. Respondent . . . Transcript on Appeal. San Francisco: Women's Co-operative Union Print, 1870, 30 pp. CLU.

5. Bar Association of San Francisco. Constitution and By-laws . . . . San Francisco: Women's Co-operative Printing Office, 1884, 20 pp. CSmH.

6. Beard, E. L., et al. In the Supreme Court . . . California . . . Appellant's Petition for a Hearing in Bank. San Francisco: Women's Print, 1880, 4 pp. CU-B.

7. Beard, Elias L., et al. In the Supreme Court . . . California . . . Transcript on Appeal . . . . San Francisco: Women's Print, 1878, 370 pp. CU-B.

Note: An unusual number of pages.

8. Beard, Elias L., et als. In the Supreme Court . . . California . . . [petition for vacating judgment] . . . . San Francisco: Women's Print, 1880, 9 pp. CU-B.

Note: Elias Beard is not mentioned in the text of this brief, only Jane M. Beard.

9. Belcher Silver Mining Co. Annual Report . . . December 31st, 1873. San Francisco: Women's Co-operative Printing Union, 1874, 19 pp. CU-B.

Note: Cover title and title page.

10. Bentley, William R. Pleasure Paths of the Pacific Northwest, Embracing the Oregon Railway . . . . San Francisco: Women's Printing Office, 1882, [64] pp. CU-B.

Note: Illustrations, plates. Cover verso has WCPU ad with woodcut which is obliterated in this copy.

11. B[lanchard], H[enry] P. A Visit to Japan in 1860 in the U. S. Frigate "Hartford" . . . . San Francisco: Women's Co-operative Printing Union, 1878, 59 pp. CU-B.

Note: Hardbound book; cover title: *At Sea and Shore.* It was issued in variant bindings (blue, green), stamped in black and gold.

12. Blum, Simon et als. In the Supreme Court . . . California . . . Appellants' Points and Authorities. San Francisco: Women's Co-operative Printing Union, 1877, 6 pp. Wendell Hammon book shop, Sacramento.

13. Boyd, James T. James T. Boyd, et al. . . . vs. Cuthbert Burrell, et al. . . . . San Francisco: Women's Print, 1880, 91, 96 pp. CU-B.

Note: 2 v. in 1; p. iii is an index. Printed on superior, heavy, white paper. This is one of the longer briefs.

14. Brooks, B[enjamin] S. Appendix to the Opening Statement . . . on the Chinese Question . . . . San Francisco: Women's Co-operative Printing Union, 1877, 160 pp. CU-B.

15. Brooks, B[enjamin] S. Brief of the Legislation and Adjudication Touching the Chinese Question . . . . San Francisco: Women's Co-operative Printing Union, 1877, 104 pp. CU-B.

16. Brown, Edgar O. In the Supreme Court . . . California . . . Transcript on Appeal. San Francisco: Women's Union Print, 1872, 20 pp. Author's Collection.

Note: This brief has no preliminary leaves. Title "Transcript on Appeal" printed in an ornamental, shaded 19th-century type.

17. Burge, Susan R. In the Supreme Court . . . California . . . Transcript on Appeal . . . . San Francisco: Women's Co-operative Printing Union, 1876, 36 pp. Author's Collection.

Note: This copy has two leaves of handwritten legal documentation tipped in after p. 36.

18. California Infantry. By-laws of the Light Guard, Company F, 1st Infantry . . . . San Francisco: Women's Co-operative Printing Union, 1878, 18 pp. CU-B.

Note: Cover title within border of oxford rule; an attractive small pamphlet.

19. California Mining Company. Annual Report . . . . San Francisco: Women's Co-operative Printing Office, 1880, 25 pp. CU-B.

20. California Mining Company. Annual Report . . . . San Francisco: Women's Co-operative Printing Office, 1881, 26 pp. CU-B.

Note: Paper is ³/₄″ smaller than 1880 ed., otherwise same format.

21. California Mutual Benefit Association. By-laws. . . . . San Francisco: Women's Union Print, 1870, 11 pp. CU-B.

Note: Cover title is within a double border, the inside border is the same as that used on all text pages, which are set in 8-pt. type.

22. California Silk Culture Association. Constitution and By-laws. San Francisco: Women's Co-operative Printing Office, 1881, 11 pp. UCB, Giannini Library.

Note: Title page with border, cover title without; six different type faces on the title pages.

23. California vs. Pablo de la Guerra. In the Supreme Court of the State of California . . . . San Francisco: Women's Co-operative Union Print, 1870, 22 pp. CHi.
Note: Cover title has rules and inner wavy border with corner pieces. Neat and crisp, with much use of small caps, italics.

24. Campbell, James. In the Supreme Court . . . California . . . Respondent's Points and Authorities. San Francisco: Women's Print, 1887, 8 pp. Wendell Hammon book shop, Sacramento.

25. Carbon River Coal Mining Company. Reports on Carbon River Gold Mines . . . . San Francisco: Women's Co-operative Printing Office, 1879, 29 pp. CU-B.
Note: Cover title and title page.

26. Casey, E. W. In the Supreme Court . . . California . . . Answer to Respondent's Brief. San Francisco: Women's Co-operative Printing Union, 1875, 6 pp. Wendell Hammon book shop, Sacramento.

27. [Caton, Amelia Z.]. One of the Cunning Men of San Francisco . . . . San Francisco: Women's Co-operative Printing Union, 1869, 56 pp. CU-B.
Note: Hardbound book; title page contains one line of Caslon caps, one line of Baskerville-like caps (with cap "i" in "Cunning" upside down) and four other type faces.

28. Chart, Obed. Memorial of Obed Chart and Others for Relinquishment of the Interest of the United States . . . . San Francisco: Women's Union Print, 1874, 16 pp. CU-B.

29. Churchill, Mrs. C[aroline] M. [Nichols]. "Little Sheaves". San Francisco: 1874, 99 pp. CSL.
Note: This copy has been rebound and the wrappers are missing and so is an imprint. There are considerably fewer pages than in the second edition (q.v.). The booklet is a collection of articles published in various papers throughout the country, possibly including *The Pioneer*.

30. Churchill, Mrs. C[aroline] M. [Nichols]. "Little Sheaves". San Francisco: Woman's Publishing Co., 1875, 136 pp. CU-B.
Note: Tan wrappers, stabbed and sewn. Cover title within ornamental border surrounded by rule border in red. Cover imprint: Woman's Pub. Co.'s Print, No. 605 Montgomery St. [Amanda M. Slocum]; title-page imprint: Woman's Publishing Co. This is a very well-printed pamphlet. Some of its contents may have originally appeared in *The Pioneer*.

31. Clark, James S. Respondent . . . Transcript on Appeal. [San Francisco: Women's Co-operative Union Print], [1870], 30 pp. CLU.

Note: Wrappers.

32. Clark, James S. In the Supreme Court of the State of California . . . Appellant . . . . San Francisco: Women's Co-operative Union Print, 1875, 11, 6 pp. CLU.

Note: Cover title.

33. Clarke, H. K. W. . . . Elizabeth Starwood, Executrix . . . vs Steiner & Klauber. . . . San Francisco: Women's Union Print, 1874, 12 pp. CU-B.

Note: Wrappers.

34. Clarke, Sarah M. Songs of Labor for the People. San Francisco: A. M. Slocum, Book and Job Printer, 1880, 17 pp. CU-B.

Note: Gray wrappers; title within double–line border.

35. Clayton, H. J. Clayton's Quaker Cook-Book . . . . San Francisco: Women's Co-operative Printing Office, 1883, 80 pp. CSmH.

Note: 24 pp. ads; WCPU ad on p.102; frontispiece a tipped-in photograph within printed rules. Glozer, *California in the Kitchen*, #65, lists as 95 pp.

36. Comstock, A. M. American Finance: Based on Gold, Silver . . . . San Francisco: Woman's Book and Job Printing Association, 8 pp. CU-B.

Note: There are no covers on this copy. Following this apparently unique imprint is the location, "605 Montgomery St." Amanda M. Slocum announced her acquisition of the Woman's [Pacific Coast] Publishing Co. and its removal to 605 Montgomery St., in the old *Evening Post* building, in February 1875. She used "Woman's Printing Association" as an imprint often thereafter, of which this appears to be a unique variant.

37. Consolidated Virginia Mining Company. Annual Report . . . . San Francisco: Women's Co-operative Printing Office, 1883, 17 pp. Author's Collection.

38. Crogha[n], Graves et al. In the Supreme Court . . . California . . . Respondents' Statement and Points. San Francisco: Women's Co-operative Printing Union, 1877, 5 pp. Wendell Hammon book shop, Sacramento.

Note: Croghan misspelled and corrected on cover in ink.

39. Currey, John. John Currey . . . vs. Juan B. Alvarado, et al. . . . Transcript on Appeal. San Francisco: Women's Co-operative

Printing Union, 1876, 140 pp. CLU.

Note: A large brief.

40. Curry, John. In the Supreme Court . . . California . . . Appellant's Points and Authorities. San Francisco: Women's Cooperative Printing Union, 1877, 9 pp. Wendell Hammon book shop, Sacramento.

41. Daly, Mary. In the Supreme Court . . . California . . . Petition for Rehearing. San Francisco: Women's Print, 1887, 23 pp. Author's Collection.

42. Dameron, J. P. et al. In the Supreme Court . . . California . . . Points and Authorities for Appellants . . . . San Francisco: Women's Print, 1885, 7 pp. CU-B.

Note: Tuttle, the attorney who wrote the brief, has his name misspelled at the end ("Tulle").

43. Dameron, Jos. P. et al. In the Supreme Court . . . California . . . Brief . . . . San Francisco: Women's Print, 1885, 4 pp. CU-B.

44. Dickenson, Harvey. In the Supreme Court . . . California . . . Harvey Dickenson, Respondent vs Joseph Gordon et als., Appellants . . . . San Francisco: Women's Co-operative Union Print, 1868, 6 pp. Not seen; item 9 in Maxwell Hunley Rare Books (Box 5151, Whittier, CA 90607) catalog, July 1980.

45. Doe, Stephen. A Lecture on Marriage . . . . 15th ed. San Francisco: Printed at the Women's Coöperative Printing Union, 1878, 16 pp. CU-B.

Note: Imprint with umlaut is correct for this title.

46. Doyle, John Thomas. Railroad Policy of California. Address . . . . San Francisco: Women's Co-operative Union, 1873, 22 pp. CSmH, CU-B.

Note: "Women's Co-operative Union" is correct imprint for this pamphlet. There was a Women's Co-operative Union at this time in San Francisco whose members were mainly engaged in the sewing trades but there was no connection between this group and the WCPU, which was devoted exclusively to printing.

47. Draymen and Teamsters' Union, San Francisco. Constitution and By-laws . . . 1876. San Francisco: Women's Co-operative Printing Office, 1879[6], 37 pp. CU-B.

Note: This copy is bound in brown buckram, a special copy with title and "Presented to the University of California" stamped in gold. The imprint date has a typographical error and should read 1876.

48. Dunbar Brothers & Co. Dunbar Brothers . . . National Bureau of Collections. San Francisco: Woman's Publishing Co., 1874, 33 pp. McCarvel (*A Check List of California Imprints* . . . *1874*) #90 assigns this title to CU-B but not found.

49. El Dorado County Historical Society. Program, Invitation, 1888. San Francisco: Woman's Printing Association, 1875, 1 pp. CHi.

Note: 1 leaf; dates are as printed. This imprint was used by Amanda M. Slocum after she acquired the Woman's Pacific Coast Publishing Company from Emily Pitts Stevens in 1875. Mrs. Slocum formed an entirely new corporation which included her own business plus her acquisition.

50. Estee, Morris M. Annual Address . . . Before the State Agricultural Society. Sacramento: Published by Order of the Society, 1874 [San Francisco: Women's Printing Union], 24 pp. CU-B.

Note: Cover title and title page; WPU imprint taken from device on verso of title page (13/16" diameter).

51. Eureka Consolidated Mining Company. Fourth Annual Report . . . . San Francisco: Woman's Publishing Company, 1874, 24 pp. CU-B.

52. Eureka Consolidated Mining Company. Eleventh Annual Report. San Francisco: Women's Co-operative Printing Office, 1880, 13 pp. CU-B.

53. Field, Stephen Johnson. Presumptions of Law in Favor of the Acts of Courts of General Jurisdiction . . . . San Francisco: Women's Union Print, 1874, 30 pp. CU-B.

Note: Cover title has no imprint.

54. [Fisher, Abby]. What Mrs. Fisher Knows about Old Southern Cooking. San Francisco: Women's Co-operative Printing Office, 1881, 72 pp. UCB, BioSci.

Note: Hardbound; Glozer, *California in the Kitchen*, #100.

55. Foresters, Ancient Order of. Constitution and By-laws . . . . San Francisco: Women's Co-operative Printing Office, 1880, 27 pp. CU-B.

Note: Collation: 24 p.+2+1.

56. Franklin, Edward . . . vs. Ephraim Merida et al. In the Supreme Court . . . California . . . Appellants' Points and Authorities. San Francisco: Women's Co-operative Printing Union, 1875, [?] pp. This title page was transcribed from Xerox copy from an unknown source.

57. Franklin, Edward . . . vs. Ephraim Merida et al. In the Supreme Court . . . California . . . Transcript on Appeal. San Francisco: Women's Co-operative Printing Union, 1875, [?] pp. Title page transcribed from Xerox copy from unknown source.

58. Franklin, Edward . . . vs. Ephraim Merida et als. In the Supreme Court . . . California . . . Appellant's [sic] Brief. Woman's Publishing Co.'s Print, 1875, [?] pp. Title page transcribed from Xerox copy from unknown source.

59. Franklin, Edward . . . vs. Ephraim Merida, Appellant. In the Supreme CourtœCalifornia . . . Transcript on Appeal. San Francisco: Woman's Publishing Co.'s Print, 1875, [?] pp. Title page transcribed from Xerox copy from unknown source.

60. Franklin, Edward . . . vs. Ephriam [sic] Merida et al. In the Supreme Court . . . California . . . Appellants in Reply. San Francisco: Woman's Co-operative Printing Union, [?] pp. Title page transcribed from Xerox copy from unknown source.

61. Fresno Canal and Irrigation Co. Findings and Decrees of the District Court, Engineer's Report . . . . San Francisco: Women's Co-operative Printing Union, 1875, 27 pp. CLU.

Note: Photocopy examined.

62. Giant Powder Company. Giant Powder, Manufactured Exclusively . . . San Francisco, Cal. San Francisco: Women's Co-operative Union Print, 1869, 25 pp. CU-B.

Note: Cover-title with imprint and title page. Style of 1869 not as crisp as later small booklets.

63. Gibbie & Barrie [publishers]. The Portfolio. San Francisco: A. M. Slocum, Printer, 188[?], 32 pp. CU-B.

Note: Gray wrappers; cover title only. Text has running titles and ruled border, with corner pieces in red. This is a catalog of engravings made from "famous paintings."

64. Gibbons, Henry. Annual Address Before the San Francisco Medical Society . . . . San Francisco: Woman's Publishing Co.'s Print, 1875, 21 pp. CSmH.

Note: Cover title only; inside title: "Modern Spiritualism."

65. Golden Gate Religious and Philosophical Society. Articles of Incorporation . . . By-laws . . . . San Francisco: Women's Co-operative Printing Office, 1886, 10 pp. CU-B.

66. Graham, Martha Morgan. An Interesting Life History . . . .

San Francisco: Women's Co-operative Printing Union, 1875, 67 pp. CLU.

Note: Essentially the same as #54.

67. Graham, Mrs. Martha M. [Stout]. The Polygamist's Victim . . . Six Years' Residence Among the Mormon Saints . . . . San Francisco: Women's Union Printing Office, 1872, 72 pp. CU-B.

68. Haley, Jane. In the Supreme Court . . . California . . . Points and Authorities for Respondent. San Francisco: Women's Print, 1883, 2 pp. Wendell Hammon book shop, Sacramento.

69. Handel and Haydn Society of San Francisco. Constitution and By-laws . . . . San Francisco: Woman's Publishing Co., 1873, 16 pp. CU-B.

Note: Cover title and title page. The title pages are from two different settings of type. The whole is well printed on good, white paper.

70. Hastings, S. Clinton. In the Supreme . . . Court California . . . Transcript on Appeal from the 12th District Court. San Francisco: Women's Co-operative Printing Union, 1875, 18 pp. Wendell Hammon book shop, Sacramento.

71. Hibberd, John F., and Piper, William A. Brief for Appellants . . . . San Francisco: Women's Print, 1882, 18 pp. CHi

72. Hills, Delia M. Whisperings of Time. San Francisco: H. Keller & Co. [Women's Print], 1878, 172 pp. CU-B.

Note: A hardbound book of poetry; it exists in at least two bindings: the first is tan cloth with blind- and gold-stamping on a cover with beveled edges and dark-brown endpapers; the second is the same but with green cloth and plain endpapers.

73. Hinkel, Charles et als. Superior Court, State of California, Appellants' Brief. San Francisco: Women's Print, 1886, 19 pp. Author's Collection.

Note: Unlike other Women's Print briefs, this has neither preliminary blank (s) nor index. Its general appearance is different from others in this listing.

74. Hoadley, Milo. In the Supreme Court . . . California . . . Brief for Respondent . . . . San Francisco: Women's Union Print, 1873, 14 pp. CHi.

75. Hoadley, Milo, Appellant. In the Supreme Court . . . California . . . Andrew Himmelman, Respondent . . . . San Francisco: Women's Co-operative Printing Office, 1872, 3 pp. CSjC.

Note: Not examined

76. Investor's Institute. Articles of Incorporation and By-Laws.

San Francisco: Women's Co-operative Printing Office, 1886, 16 pp. CU-B.

Note: Stiff, dark-blue covers with bronzed lettering. Included in the imprint is the new address, 23 First Street, where the firm had joined Hicks-Judd, bookbinders and printers.

77. Journeymen Shipwrights' Association of San Francisco. Constitution and By-laws . . . . San Francisco: Women's Union Print, 1869, 16 pp. CU-B.

Note: Cover title and title page with imprint.

78. Keystone Quartz Mining Company. In the Supreme Court . . . California . . . Brief . . . . San Francisco: Women's Co-operative Union Print, 1871, 22 pp.

Note: Title "from an old 'Western Hemisphere' bookseller's catalog." Citation furnished by Allan R. Ottley.

79. Lake. The Dark Seance, a Farce in Two Acts. San Francisco: Women's Union Print, 1872, 17 pp. CU-B.

Note: Gray Wrappers, cover title within a border and title page. The imprint appears only on p. 17 and is in the form of a logo. This is apparently a Spiritualist spoof. Entered for copyright purposes by H. L. Knight who may or may not have been person using name of "Lake."

80. Larrabee, Charles H. In the Matter of the Survey of the Ranchos . . . Brief . . . . . San Francisco: Women's Co-operative Union Print, 1869, 60 pp. CU-B.

81. Lazard, E. et al. In the Supreme Court . . . California . . . Transcript on Appeal. San Francisco: Women's Co-operative Printing Union, 1875, 8 pp. Author's collection.

Note: The brief deals with partnership of Lazard Frères, a famous financial firm founded in San Francisco. It is printed on watermarked, laid bond of a superior quality.

82. Lazard, E. et als. In the Supreme Court . . . California . . . Appellant's Points and Authorities. San Francisco: Women's Co-operative Printing Union, 1875, 6 pp. Wendell Hammon book shop, Sacramento.

83. Liés, Eugene. . . . Pablo de la Guerra . . . vs Eugene L. Sullivan . . . . San Francisco: Women's Union Print, 1872, 12 pp. CU-B.

84. Likins, Mrs. J. W. Six Years Experience as a Book Agent in California . . . . San Francisco: Women's Union Book and Job Printing Office, 1874, 168 pp. CU-B.

Note: Book was issued in wrappers with cover title imprint "Women's print"; title page has unique, full imprint. Apostrophe missing in all versions of the title. It was reprinted by The Book Club of California in 1992.

85. Lorenz, Henry. In the Supreme Court . . . California . . . Brief for Respondent. . . . San Francisco: Women's Print, 1884, 9 pp. CU-B.

86. Lucas & Company. Gypsum as a Fertilizer . . . . San Francisco: Women's Co-operative Printing Office, 1886, 18 pp. CU-B.

Note: Light-green, laid cover with title in red; l p. ads. Display types bad and underlined.

87. Lyle, Robert. In the Supreme Court . . . California . . . Brief. San Francisco: Women's Print, 1884, 21 pp. CU-B.

88. Mahony, Mary G. Marmaduke Denver and Other Stories. San Francisco: Women's Co-operative Printing Office, 1887, [vi], 119 pp. CU-B.

Note: Hardbound book with patterned endpapers.

89. Mardwell, C. F. Illustrated Catalogue and Price List of Tools & Supplies. San Francisco: Jennie E. Patrick Printing Co., 413-415 Market St., 1888, [?] pp. CHi.

90. Martin White Mining Co. Brief of Respondent . . . . San Francisco: Women's Print, 62 pp.

Note: Item #974 in Robert D. Armstrong, *Nevada Printing History, 1858–1880* (Reno: University of Nevada Press, 1981), p. 326. Not seen.

91. McDermott, A., et al. In the Supreme Court . . . California . . . Petition for Rehearing . . . . San Francisco: Woman's Publishing Co., 1874, 11 pp. Author's Collection.

92. McLeran, Thomas G., vs. Benton, J. E. et al. In the Supreme Court . . . California . . . Transcript on Appeal . . . . San Francisco: Women's Co-operative Printing Union, 1875, 128 pp. CLU.

93. McNamee, Andrew, vs. McCusker, Daniel et al. In the Supreme Court . . . California . . . Transcript on Appeal. San Francisco: Women's Union Print, 1874, i, ii, 65 pp. CLU.

Note: Pp. i, ii are index. Poor presswork, inking.

94. Meadow Valley Mining Company. By-laws . . . . San Francisco: Women's Union Print, 1870, 15 pp. CU-B.

Note: Small format with cover title and title page and double rules on pages. Good paper; work compares favorably with similar items by Bosqui, Eastman, etc.

95. Mechanics' Institute. Report of the Sixth Industrial Exposition . . . . San Francisco: Women's Co-operative Union Print, 1868 [1869], 94 pp. CU-B.

Note: Title page imprint is 1868, cover 1869. This report contains much small type and many tables.

96. Mechanics' Institute. Report of the Seventh Industrial Exposition . . . . San Francisco: Women's Co-operative Print, 1870, 102 pp. CU-B.

Note: Cover title only in red and blue with two-color border; contains woodcut of pavilion.

97. Mendelssohn Quintette Club of Boston. Thirty-first Season . . . . San Francisco: Women's Print, 1880, 15 pp. CU-B.

98. Moerenhout, Jacob A., vs. Williams, Henry F. In the Supreme Court . . . California . . . . Transcript on Appeal. . . . . San Francisco: Women's Co-operative Print, [1870?], 131 pp. CLU.

99. Monterey, [City of]. Claims of Monterey, California, for the New Soldiers' Home. San Francisco: Women's Print, [1887?], ii, 16 pp. CSmH.

Note: Contains map and engravings; pp. i and ii are ads; Women's print, 23 First St., on cover title only; uses Mackellar, Smiths & Jordan's oak-leaf border.

100. [National Labor Union]. Address of the Executive Committee . . . . San Francisco: Women's Co-operative Printing Union, 1871, 72 pp. CU, CSmH.

Note: Purple wrappers with cover title and title page.

101. [Nevada petition]. . . . A Protest and Petition . . . Against Unlawful Practices at the Carson City Branch Mint . . . . San Francisco: Women's Union Print, 1873, 60 pp.

Note: Gray wrappers, cover title and title page. Item 728 in Robert D. Armstrong, *Nevada Printing History* . . . 1858–1880 (Reno: University of Nevada Press, 1981), p. 249. Title page is a 12-line variant of cover title. There are two appendixes, each with separate title page and numbered at the top continuously with the main text. Each appendix also has separate pagination at the bottom of the page. There are errata slips tipped in on p. 29 of first appendix and on p. 37 of the second.

102. Nunan, Matthew, and Smith, N. Proctor [appellants]. In the Supreme Court . . . California . . . Transcript on Appeal . . . . San Francisco: Women's Print, 187[9?], 17 pp. CU-B.

103. Olympic Club of San Francisco . . . . Twentieth Annual Exhibition. . . . San Francisco: Women's Print, 1880, 8 pp. CU-B.

Note: Bronzed border, type red, all on blue stock; eleven different types on p.1; 4 pp. ads.

104. Onderdonk, A. In the Supreme Court . . . California . . . Points and Authorities for Respondent. San Francisco: Women's Print, 1888, 12 pp. Wendell Hammon book shop, Sacramento.

105. Packard, Albert. Argument . . . of the Heirs of Caesario Lataillade . . . . San Francisco: Women's Union Print, 1873, 15 pp. CU-B.

Note: Not located as of 1/23/91.

106. Panamint Mining and Concentration Works. Prospectus . . . . San Francisco: Women's Print, [1875?], 7 pp. CU-B.

Note: Frontispiece a colored folding plate.

107. Peckham, Mrs. P. Annetta. Cuttings . . . . San Francisco: Amanda M. Slocum, Book and Job Printer, 1877, 71 pp. CU-B.

Note: Frontispiece a Bradley & Rulofson mounted photograph, guarded-in; press notices, pp. 70-71.

108. People of the State of California . In the Supreme Court . . . California . . . Appellants' Points and Authorities . . . . San Francisco: A. M. Slocum, Book and Job Printer, 1880, 4 pp. CU-B.

109. People of the State of California [appellants]. In the Supreme Court . . . California . . . Transcript on Appeal . . . . San Francisco: Women's Print, 1878, 12 pp. CU-B.

110. Pilkington, J. Religion and Science . . . . San Francisco: Woman's Publishing Co., 1875, 32 pp. CU-B.

Note: Cover title and title page. Large type and neatly done; better than most religious printing, Murdock's excepted, of this period.

111. Port Paraiso Lead and Silver Mining Company. Prospectus and By-laws . . . . San Francisco: Women's Co-operative Union Print, 1873, 12 pp. CU-B.

Note: Cover title and title page; seven different types in nine lines of title; double-hairline rules around every page; small format.

112. Porter, George K. In the Supreme Court . . . California . . . Transcript on Appeal. San Francisco: Women's Co-operative Printing Union, 1875, 89 pp. Wendell Hammon book shop, Sacramento.

Note: One of larger briefs.

113. [Presbyterian Synod of Alta California]. Minutes of . . .

Annual Meeting of 1868 . . . . San Francisco: Women's Co-operative Union Print, 1868, 16 pp. WPA Imprints Inventory, 1868; Drury, *California Imprints, 1846–1876*; not seen.

114. Ramsdell, Sarah A. Backward Glimpses Given to the World by John Bunyan . . . . San Francisco: Woman's Publishing Company, 1873, 165 pp. CU-B.

Note: A hardbound book about Spiritualism.

115. Ranchos Cuyama . . . . In the Matter of the Contested Survey . . . . San Francisco: Women's Union Print, 1873, 11 pp. CU-B.

Note: Ink correction of transposed "and" on p.11.

116. Reay, Joseph W. et als. In the Supreme Court . . . California . . . Transcript on Appeal. San Francisco: Women's Print, 1880, 39 pp. Author's Collection.

Note: Index is on p. [iii] and is a cancel tipped to conjugate blank, [i], [ii].

117. Redwood Lumber Association. Certificate of Incorporation and By-laws . . . . San Francisco: Women's Union Print, 1871, 16 pp. CU-B.

Note: Some double-hairline rules as with other by-laws, etc.

118. Reimer, Ellen. In the Supreme Court . . . California . . . Points, Etc., on Part of Respondent. San Francisco: Woman's Printing Association, 1875, 6 pp. Author's Collection.

Note: This brief has no preliminary blank; the text begins on p. [1].

119. Republican State Central Committee of California]. The Record of Governor Haight. [San Francisco: Women's Co-operative Print], 187[1?], 16 pp. CLU.

Note: Self-cover; imprint is on p.16.

120. Reynolds, Rachel Ann . . . . . In the Supreme Court . . . . February 1994 . . . California . . . Transcript on Appeal. San Francisco: Women's Co-operative Printing Union, 1877, 23 pp. Author's Collection.

Note: Despite the use of six type faces on the cover title, including Caslon, the results are not cluttered or unattractive.

121. Reynolds, Rix & Co. Dry Air Compressors . . . . San Francisco: Women's Print, 1876, 40 pp. CU-B.

Note: Illustrated with many wood engravings of New York origin. Contains much complicated typesetting.

122. Rhoda, Frederick [et al.]. In the Supreme Court . . . California . . . Petition for Re-Hearing. San Francisco: Women's Co-operative Printing Union, 1877, 10 pp. Wendell Hammon book shop, Sacramento.

123. Rhoda, Frederick [et al.]. In the Supreme Court . . . California . . . Transcript on Appeal. San Francisco: Women's Print, 1882, 15 pp. Wendell Hammon book shop, Sacramento.

124. Richmond, Mrs. Cora L. V. The Nature of Spiritual Existence . . . . San Francisco: Women's Co-operative Printing Office, 1884, iv, 172 pp. CU-B.

Note: Spiritualist discourses given through the mediumship of Mrs. Cora V. Richmond at San Francisco, California, in 1883.

125. Riggers' and Stevedores' Union Association. Revised Copy of the Constitution and By-laws . . . . San Francisco: Women's Co-operative Print, 1878, 23 pp. CU-B.

Note: Illustrated; a wood-engraving, signed "Burr, San Francisco," is used as tailpiece between constitution and by-laws sections.

126. [Rogers, Ada B.]. Cousin Phebe's Chats with Children. San Francisco: Women's Co-op. Printing Office, [1885], 72 pp. CU-B.

Note: Frontispiece [iv], plates; imprint taken from verso of title page; bound in green cloth with ornamental black stamping, gold title and vignette. Includes five wood engravings, one being repeated as frontispiece. All the engravings are mortised indicating a former use (unknown). Each story begins with a fancy drop initial, and there are typecast, pen-style ornaments as tailpieces. This is a well-printed book.

127. Root, Julia Anderson. Healing Power of Mind . . . . San Francisco: Women's Co-operative Printing Office, 1884, xii, 157 pp. CU-B.

Note: Hardbound in blind-stamped, blue cloth with title and author in gold.

128. Royer, Herman. In the Supreme Court . . . California . . . Appellant's Brief. San Francisco: Women's Print, 1880, 30 pp. Wendell Hammon book shop, Sacramento.

129. Royer, Herman. In the Supreme Court . . . California . . . Transcript on Appeal. San Francisco: Women's Print, 1884, 60 pp. Wendell Hammon book shop, Sacramento.

130. San Diego [City of]. . . . City of San Diego . . . vs. Robert Allison . . . Brief of Appellant . . . . San Francisco: Law Printing House of Woman's Publishing Co., 1873, 4 pp. CHi Pam 507.

Note: Not seen.

131. San Felipe Mining Co. . . . vs. M. S. Belshaw. San Francisco: Women's Union Print, 1873, 25 pp. CHi.
Note: Cover title within three-rule borders; generous margins throughout.

132. San Francisco Gas Light Co. In the Supreme Court . . . California . . . Brief for the Defendant. San Francisco: A. M. Slocum, Printer, 1879, 5 pp. CU-B.

133. [San Francisco] Ladies' Protection and Relief Society. Twenty-first and Twenty-second Annual Reports . . . . San Francisco: Women's Co-operative Print, 1875, 27 pp. CU-B.
Note: Cover title and title page.

134. San Francisco Port Authority. Proceedings of the Eleventh Anniversary . . . . San Francisco: Women's Co-operative Union, 1871, 16 pp. CU-B.

135. San Francisco Stock and Exchange Board. Constitution and By-laws . . . . San Francisco: Women's Union Print, 1874, 29 pp. CU-B.
Note: Cover title and title page; oxford rules on cover only; six types on title page. Good margins.

136. San Joaquin and King's River Canal and Irrigation Company. Report . . . for 1873 . . . . San Francisco: Woman's Publishing Company, 1874, 38 pp. CU-B.
Note: Folding map missing from both CU-B copies.

137. Sayward, W. T. All about Southern California . . . . San Francisco: Woman's Publishing Co.'s Print, 1875, 20 pp. CU-B.
Note: The printer was at 605 Montgomery St. at this time.

138. Schultz, John and Henry von Bergen. In the Supreme Court . . . California . . . Transcript on Appeal. San Francisco: Women's Print, 1885, 72 pp. Author's Collection.
Note: This is a substantial brief.

139. [Scott, William Anderson]. In Memoriam of Thomas Breeze . . . . San Francisco: Women's Co-operative Printing Union, 1874, 10 pp. Source of this imprint unknown.

140. Shields et al. In the Supreme Court . . . California . . . Appellant's Points and Authorities. San Francisco: Women's Print, 1883, 6 pp. Author's collection.

141. Sierra Nevada Mine. Annual Reports . . . . San Francisco: Women's Print, 1880, 2 pp. CU-B.
Note: The legal-size paper is printed the long way of the sheet. Clear type and tables.

142. Slocum, Amanda M., and W. N., eds. *Common Sense. A Journal of Live Ideas.* San Francisco: Woman's Publishing Co., 1874–1875 [2 v. in 1], 626; 28 pp. CU-B.

Note: Includes material on Spiritualism, suffrage, etc. Publication ceased with second issue of v. 2.

143. Society of California Pioneers. Report of the Committee ... on the Financial Condition .... San Francisco: Women's Co-operative Printing Union, 1878, 8, [12] pp. CU-B.

Note: Cover title and title page; [12] pp. are tables, in landscape format with other, smaller, tables within the text. All are accurately composed and clear.

144. Spivalo, Aug. D. et al. In the Supreme Court ... California ... Transcript on Appeal .... San Francisco: Women's Print, 1881, 119 pp. CU-B.

Note: Neatly done and comparable to those briefs produced by contemporary San Francisco printers. Note length.

145. [St. John, Brotherhood of]. Articles of Incorporation and By-laws .... San Francisco: Women's Co-operative Printing Office, 19 pp. CSmH.

Note: Gray cover with title within oxford rules; title page.

146. Stewart, George W., et als. In the Supreme Court ... California ... Respondents' Points. San Francisco: Women's Print, 1887, 5 pp. Wendell Hammon book shop, Sacramento.

147. Stratton, Francis E. In the Supreme Court ... California ... Appellant's Points & Authorities. San Francisco: Women's Print, 1885, 10 pp. Wendell Hammon book shop, Sacramento.

148. Stretch, R. H. Report on the Amador Canal and Mining Company .... San Francisco: Women's Co-operative Printing Office, 1879, 57 pp. CU-B.

Note: Cover title and title page. "Note" tipped in before p. [3].

149. Stretch, Richard H. Report on the Amador Canal and Mining Co. San Francisco: Women's Co-operative Printing Office, 1880, 48 pp. CU-B.

Note: Gray wrappers with cover title and title page; addendum between cover title and title page.

150. Stretch, R. H. Report on the Amador Canal and Mining Company. San Francisco: Women's Co-operative Printing Office, 1880, 48 pp. CU-B.

Note: Gray wrappers with cover title and title page; addendum tipped between cover and p. [1].

151. Studebaker Bros Mfg. Co. Catalogue . . . . San Francisco: Women's Co-operative Printing Office, [1883?], 96 pp. CHi.
Note: Cover title and title page; landscape format with much tabular matter and many wood-engravings. Neatly printed.

152. Sullivan, Eugene L. In the Supreme Court . . . California . . . Respondent's Points and Authorities. San Francisco: Women's Co-operative Printing Union, 1877, 8 pp. Wendell Hammon book shop, Sacramento.

153. Sutro Tunnel Co. Annual Report . . . . San Francisco: Women's Co-operative Printing Office, 1881, 23 pp. CU-B.
Note: Beige wrappers, cover title and title page; folding map.

154. Sutro Tunnel Co. Annual Report . . . . San Francisco: Women's Co-operative Printing office, 1882, 19 pp. CU-B.
Note: Cream wrappers, cover title and title page; folding map.

155. Sutro Tunnel Co. Annual Report . . . . San Francisco: Women's Co-operative Printing Office, 1883, 13 pp. CU-B.
Note: Cover title and title page; folding map.

156. Sutro Tunnel Co. Annual Report . . . . San Francisco: Women's Co-operative Printing Office, 1885, 11 pp. CU-B.
Note: Gray wrappers, cover title and title page; folding map.

157. Sutro Tunnel Co. Certificate of Incorporation and By-laws. San Francisco: Women's Co-operative Printing Office, 1880, 16 pp. CHi.
Note: Cover title and title page same except cover title has rules, as was often the practice at the WCPU.

158. Thompson, Francisco. In the Supreme Court . . . California . . . Appellant's Brief. San Francisco: Women's Co-operative Printing Union, 13 pp. Wendell Hammon book shop, Sacramento.

159. Thorndyke, Mrs. E. P. Astrea, or Goddess of Justice. San Francisco: Amanda M. Slocum, Book and Job Printer, 1881, 106 pp. CU-B.
Note: Hardbound book (this copy: green cloth). Published at 612 Clay St.

160. Thorne, Susan E. et al. In the Supreme Court . . . California . . . Defendant's Points and Authorities. San Francisco: Women's Print, 1885, [?] pp. Wendell Hammon book shop, Sacramento.

161. Tibbets, S. M., and Simpson, A. M. [counsel]. Memorial . . . for Establishment of Southern Line of Presidio Reservation

. . . . San Francisco: Women's Union Print, 1874, 18 pp. CU-B.
Note: Folding maps.

162. [Traveler's Guide]. . . . Traveler's Guide for California and the Pacific Coast. San Francisco: Women's Co-operative Printing Union, July, Aug. 1869, [32], [32] pp. CU-B.
Note: Wrappers, cover title. There is no pagination but the quarto signatures are numbered 1–4. These are compendiums of vessel, railroad and stage timetables; printing is in red, blue and black.

163. U. S. District Court, Calif. (Northern District). Decision of Hon. Ogden Hoffman . . . . San Francisco: Woman's Publishing Co., 1874, 14 pp. CLU.

164. U. S. Land Office. . . . In the Matter of . . . the Keystone Consolidated Mining Company . . . . San Francisco: Women's Union Print, 1871, 12 pp. Author's Collection.

165. U. S. General Land Office. Report of Commissioner Upon the Charges of Official Misconduct . . . . San Francisco: Women's Co-op. Union Print, 1874, 44 pp. Source of this imprint unknown.

166. [United States]. Act of Congress Creating the Office of Shipping Commissioner . . . . San Francisco: Women's Print, 1872, ii, 20 pp. CU-B.
Note: Cover title is within fancy, non-right-angle corners that match oxford rules.

167. [United States]. Act of Congress Creating the Office of Shipping Commissioner. San Francisco: Women's Co-operative Print, 1873, ii, 20 pp. CU-B.
Note: Index on p. [i].

168. [United States]. Revised Statutes . . . Relating to Merchant Seamen . . . . San Francisco: Woman's Publishing Co.'s Print, 1874, [ii], 49 pp. CU-B.
Note: Light inking on heavy stock (good, white paper); this pamphlet has shoulder notes and notable margins as well.

169. Wagner, Theodore. . . . In the U. S. Land Office at San Francisco . . . James P. Hamilton. San Francisco: Women's Print, 10 pp. CU-B.

170. Wagner, Theodore. In the Dept. of the Interior . . . Christian vs. Strentzel . . . Brief . . . . San Francisco: Women's Co-operative Printing Office, 1886, 22 pp. CU-B.

171. Wagner, Theodore. In the Dept. of the Interior . . . [Sassenberg vs. Gleason] Brief . . . . San Francisco: Women's Co-operative Printing Office, 1886, 15 pp. CU-B.

Note: The WCPU was at 23 First St. by this time.

172. Wagner, Theodore. . . . In the Matter of H. G. F. Dohrman . . . vs. William Krieger . . . . San Francisco: Women's Print, 17 pp. CU-B.

173. [Washington, George]. Farewell Address . . . . San Francisco: Women's Co-operative Printing Union, 1881, 23 pp. Author's Collection.

Note: Wrappers with cover title only. "Printed for the American Union," not otherwise identified.

174. Weed, Joseph. Recollections of a Good Man . . . . San Francisco: A. M. Slocum, Book and Job Printer, 1880, 17 pp. CU-B.

Note: Gray wrappers. Cover title within border followed by a blank not in pagination; title page.

175. Wells, Fargo & Co. By-laws . . . . San Francisco: Women's Co-operative Printing Union, 1873, 6 pp. CSmH.

176. Williams, James. Life and Adventures of . . . a Fugitive Slave, with a Full Description of the Underground Railroad. San Francisco: Women's Union Print, 1873, 108 pp. CSmH, CSL, CU-B.

Note: This is the second edition; the first was a Sacramento imprint (1873) with 48 pp.; map missing from CU–B copy. CSmH copy in brown wrappers with cover title. Preface signed: "John Thomas Evans (formerly) now James Williams [b. 1825]."

177. Williams, James. Life and Adventures of . . . a Fugitive Slave, with a Full Description of the Underground Railroad. San Francisco: Women's Union Print, 1874, 124 pp. CU-B.

Note: Third edition, with sixteen additional pages. Orange wrappers; cover title. "This book was begun in 1869, and printed in 1873. I, James Williams, commenced to sell pamphlets of my book in April, 1853" (p. 104). "He will remain in town a day or two, and call upon our people, so that they can, for 50 cents, procure a copy of his book, and help a poor man" — *Yolo Democrat* (p. 123).

178. Wilson, John D. et al. In the Supreme Court . . . California . . . Transcript on Appeal. San Francisco: Women's Print, 1884, 12 pp. Author's Collection.

Note: Instead of being centered within the outer two rules and the inner rule, the imprint is flush left, a typographical error.

179. Young, Carrie F., ed. *Woman's Pacific Coast Journal.* San Francisco: Women's Co-operative Print, [May–Dec.] 1870., 128 pp. CU-B.
Note: Imprint varies from none (v. 1, #3) to "Publication Office, 424 Montgomery St." (v. 1, #4) and "Women's Co-operative Printing Union, at Publication office, 424 Montgomery St." (v. 1, #8). Eight issues only were printed at the WCPU.

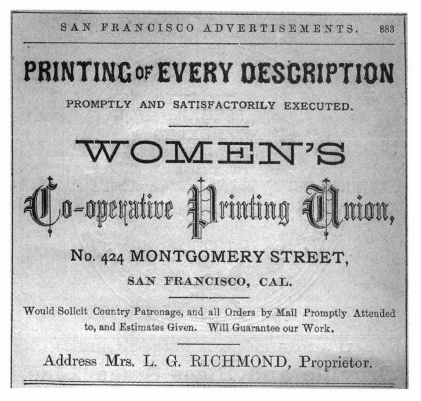

Fig. 42: *This is an advertisement from Langley's 1878 Directory.* Author's Collection.

Fig. 43, following page: *An unusual example of a handbill showing the kinds of nostrums being peddled in the 19th Century. Last line reads: "Women's Print, 424 Montg'y St. S. F."* Author's Collection.

DR. LEVINGS'

SYRUP OF

HOARHOUND

AND

ELECAMPANE

FOR THE CURE OF

Pneumonia, Coughs, Colds, Croup, Asthma, Catarrh, Whooping Cough,

AND BRONCHIAL AFFECTIONS.

This celebrated Medicine was formerly put up in the City of New York by Dr. Levings, and upon its own merits obtained for itself a large demand without the usual method of publication and advertising.   As a reliable and safe

FAMILY MEDICINE

I am satisfied it has never been excelled.   Four of the ingredients in its composition are known to every housewife in the country; though not the most active remedies in the preparation, they are good so far as they go.   Though a Vegetable Preparation, it is a very active Medicine.   It will cause easy expectoration in from 30 minutes to one hour after it is taken.   If you are troubled with

COLD FEET,

By taking a dessert spoonful on going to bed, in one hour you will be in a warm glow all over the system.   If you have a very

Obstinate Cold,

Take, in the afternoon, as directed on the bottle; then at night take a dessert spoon-ful, and also a warm foot bath upon going to bed, and I have never known it to fail to break up a cold in one or two nights.

In Case of Croup,

Bathe the child's feet in warm water for five minutes, then wrap them up warm; give the Syrup as directed on the bottle for children.

Whooping Cough

Is cured in three or four days by the use of this Syrup, if taken in the first stages of the complaint.   It should be taken in doses of a teaspoonful each, through the day, for the cure of Asthma and Bronchial Affections.

Every family having children should not be without a bottle or two of this syrup in the house, in case children are attacked with Croup.   It is a safe and reliable Remedy.

Sold by the case by all wholesale Druggists and dealers in medicine.

Women a Print, 424 Montg'y St. S. F

APPENDIX C

MALE PRINTERS IN FEMALE PRINTING OFFICES
Northern California, 1857–1890

1. Fish, Edward P. 928 Clay St. (SF). Compositor, WCPU.
SOURCE: 1871 Directory.

Note: Fish later became foreman at Cubery & Co.

2. Fish, Milo H. 121 8th St. (SF) (2)(3). 1 Polk St. (4). 2506 Post
St. (5). 2526 Post St., rear (6). 2516 Post St. (7). 2506 Post St. (8).
2514 Post St. (9). Pressman, Carrier Dove Printing & Publishing Co. (1). Printer (2)(3)(7). Schlesinger & Fish [printers] (4)
(5). Conductor (6). Driver, A. W. Weldon ["Carrier, Post Office"] (8)(9). SOURCES: (1) 1890 Directory. (2) 1891 Directory.
(3) 1893 Directory. (4) 1894 Directory. (5) 1895 Directory. (5)
1895, 1896, 1897 Directories. (6) 1898 Directory. (7) 1899, 1900
Directories. (8) 1901 Directory. (9) 1902, 1903 Directories.

Note: The Carrier Dove Printing & Publishing Co. was a family business.
Fish married Nellie Gorman (q.v.), a typesetter who evidently came over to
the new printing-office from the WCPU where the magazine had been formerly printed. In 1894, the printing business became "Schlesinger & Fish,
proprietors, Carrier Dove Printing Co.," 1231 Market St. (Milo Fish was
stepson to Louis Schlesinger.) In 1899, after the *Carrier Dove* shop had been
closed long since because of two separate fires, Milo Fish is in the Directory under "Printers — Book & Job" as being at 538 California St. with James

M. Blakely in what was probably a short-lived partnership for neither is listed there again.

3. Gallagher, George. 826 1/2 Harrison St. (SF). Pressman, WCPU. SOURCE: 1871 Directory.

Note: The 1872 Directory shows him apprenticed to the Fulton Foundry and by 1873 he is a machinist with that firm. Henry G. Gallagher, at the same Harrison St. address above, is listed as a pressman at Carmany & Co.

4. King, George C. 1111 Mason St. (SF) (1). 1520 California St. (2). 1520 Clay St. (4). 1322 Clay St. (5). Printer, Bosqui Printing & Engraving Co. (1). Foreman, History Department, A. L. Bancroft & Co. (2). Foreman, WCPU (3). Printer (4). Compositor, Palmer & Rey (5). SOURCES: (1) 1882 Directory. (2) 1883 Directory. (3) *Pacific Printer,* May 1884, p. 11:1. (4) 1884, 1885 Directories. (5) 1886, 1887 Directories.

Note: The full entry from *Pacific Printer:* "George King, formerly of Bosqui & Co., is now foreman for the Women's Co-operative office."

5. Oetzel, John H. 207 Post St. Pressman, WCPU. 1869 Directory.

6. Richmond, Isaac L. 518 William St. (O) (1)(3)(4). N side William near Telegraph (2). Compositor, WCPU (1). Compositor, SF (2). Stationery (3). Clerk (4). SOURCES: (1) 1878 Bishop Directory. (2) 1878 Directory (O). (3) 1878 Bishop Directory (O). (4) 1879 Bishop Directory (O).

Note: Isaac Lester Richmond, younger son of Lizzie G. (q.v.), died 6 April 1879 when he was only 17 years old, according to cemetery records.

7. Richmond, Willard P. Richmond, Willard P. and Mrs. Mary H. (22). 726 California St. (SF) (1). N side William near Telegraph (O) (2)(4)(6). Oakland (3)(5)(7)(9)(11)(15). 518 William St. (O)(8)(10)(12)(14)(16)(17)(18)(25). 807 California St (SF) (19). 1517 Vallejo St. (SF) (20)(21). 514 Stockton St. (SF) (22). 1613 Jones St. (SF) (23)(24). 309 Broadway (O) (26). 1301 Clay St. (O) (27). Compositor (WCPU)(1)(3)(5)(7)(9)(10)(20). Printer (2)(8)(12)(21)(24). (Mrs.) L. G. Richmond & Son (SF) (4)(6) (11)(15)(16)(17)(18). Printing-office, SF (14). Willard P. Richmond & Co., Job Printers (22)(23). Pacific Coast Agent, Physicians' and Surgeons' Soap (25). Compositor, *Enquirer* (26)(27). Age: 24. Born: Rhode Island (13). SOURCES: (1) 1874, 1875 Directories. (2) 1875 Directory (O). (3) 1876 Directory. (4) 1876 Directory (O). (5) 1877 Bishop Directory. (6) 1877 Directory (O). (7) 1878 Directory. (8) 1878 Directory (O), 1878 Bishop Directory (O). (9) 1879 Directory. (10) 1879 Bishop Directory (O). (11) 1880 Directory. (12) 1880, 1881 Directory (O). (13) 1880 Cenus. (14) 1883 Directory (O). (15) 1881–1888 Directories. (16) 1884 Directory (O). (17) 1885, 1886 Directories (O). (18) 1887 Directory (O). (19) 1888 Directory. (20) 1889 Directory. (21) 1890, 1891 Directories. (22) 1893 Directory. (23) 1894 Directory. (24) 1895 Directory. (25) 1896 Directory (O). (26) 1898 Directory (O). (27) 1899 Directory (O).

Note: After the WCPU was sold to the Hicks-Judd Co., Willard Richmond remained as a principal with the newly incorporated firm. Occasionally, he was erroneously listed in a Directory as "William" or his street address incorrectly given as "518 Williams St."

8. Schlesinger, Louis. 32 Ellis St. (1). Printer, 841 Market St. (1). Manager, Carrier Dove Printing and Publishing Co. (2) Co-publisher, *Carrier Dove* (3). Schlesinger & Fish Printing Co. (4). SOURCES: (1) 1888 Directory. (2) 1890–1893 Directories. (3) Masthead, *Carrier Dove*, passim. (4) 1894, 1895 Directories.

Note: The Carrier Dove Printing and Publishing Co. was originally at 841 Market St., later at 1231 Market St. After the close of the *Carrier Dove* in 1893, Schlesinger and his stepson, Milo H. Fish, continued their work in printing at 1 Polk St. and then 534 Page St. Prior to 1888, Louis Schlesinger was usually listed in Directories as "medium and magnetic healer" and eventually he added "M.D." to his titles.

9. Sprague, George. Compositor, *Pacific Monthly*. SOURCE: 1864 Directory.

Note: Sprague was an officer in the the short-lived Female Typographical Union announced by Lisle Lester in May 1864.

10. Warren, Patrick J. 59 Minna St. (SF) (1). 435 Pine St. (2). Printer (1). Pressman, Women's Co-operative Union [sic] (2). SOURCES: (1) 1870 Directory. (2) 1872 Directory.

11. Wickes, Frank [S.]. 803 Hyde St. (SF) (1) 14 Geary St. (SF) (2). 314 Bush St. (3). Foreman compositor, *Pioneer* (1). Compositor, *Pioneer* (2) Printer, Woman's Publishing Co (3). SOURCES: (1) 1871 Directory. (2) 1872 Directory. (3) 1873 Directory.

Note: Wickes began his local career as a writer for, and then co-proprietor and co-editor of, the *Saturday Evening Mercury* which he and Mrs. Emily Pitts (Stevens) took over in January 1869. She bought him out in November and he then became a printing employee and remained with Mrs. Pitts Stevens when she started her newly incorporated Woman's Publishing Co. in 1872.

INDEX
By Vivian Fisher